How to Buy Design

MARION HANCOCK

The Design Council

Published in the United Kingdom in 1992 by

The Design Council

28 Haymarket

London SW1Y 4SU

Printed and bound in the United Kingdom by

Bourne Press Ltd, Bournemouth

Designed by Spy Design, London

Edited by Victoria Felton and Tracey Williams

Every effort has been made to check the factual accuracy of the information given in this book, at the time of publication. The views and opinions expressed are those of the author and do not necessarily reflect or represent the views and opinions of the Design Council.

British Library Cataloguing in Publication Data.

A catalogue record for this book is available from the British Library.

ISBN 0 85072 301 9

CONTENTS

Marion Hancock is the editor of *Design*,
the international magazine published
monthly by the Design Council for
designers and their clients. In this and
her last job, as an editor and project
manager for the British Council, she has
worked regularly with design
consultancies; she has also researched
successful and unsuccessful design
projects of all types.

PREFACE

'These guys are between a rock and a hard place. They got given responsibility for the design project because they were the last ones into the boardroom. They don't even want the project, but their future career may depend on their handling it successfully.'

Keith Lawson
Business Design Group

This book is for people who have acquired the responsibility for a design project; people who need to use design for business purposes. Specifically, the book deals with the practicalities of using design consultancies: where to find them, how to choose between them, how much their services will cost, how to put together a brief, and what input they will need from you. Whether the project involves graphic design, product design, interior design or a combination of all three, the mechanics of commissioning a design consultancy are much the same.

In breaking the process down into step-by-step units, I am aware I have not really done justice to the satisfaction to be gained from working with designers and seeing imaginative and apt solutions to business problems emerge. Although all design projects encounter problems, using design consultancies can also be very enjoyable. I hope that this book may help the unwary to maximize the satisfaction and minimize the problems.

I also hope that it may serve as an introduction to other publications which deal in more detail with specific types of design project. I have mentioned some of these in the Publications section.

I would like to thank everyone whose opinions have contributed to this book. In particular, I would like to thank the staff of the Design Council's Designer Selection Service, Peter Tennent, Roy Fleming and Nicky Hough; also Michael Zur-Spiro, Maxine Horn, Roger Cooper, Steve Hallam, Keith Lawson, Robin Levien and Caroline Middleton.

Marion Hancock

February 1992

1 benefits

1 REDUCING RISK

Many businesspeople regard dealing with design as a chore; expensive, tedious and, above all, risky. Designers suffer, for their part, from association with what *The Financial Times* described in 1991 as gratuitous 'logo-mongering and shop-revamping'. Undoubtedly this reputation is sometimes deserved. The classic gripe among people who have had bad experiences with designers is that 'The designer draws lots of pretty pictures, walks off and leaves the client holding the baby – then the baby doesn't work'.

Clients themselves are often blind to their own contribution to this state of affairs. A 1990 survey conducted by the Open University together with the University of Manchester Institute of Science and Technology (UMIST), *Benefits and Costs of Investment in Design*, looked at the outcome of investment in design across a broad range of design projects, spanning shoes, textiles, graphics, packaging, furniture and kitchenware. Ten per cent of these projects failed to produce a usable result. It was found, however, that the unsuccessful projects failed because of problems in managing the design consultants, and in particular because of inadequate briefing. This was especially true of the smallest firms.

In the other 90 per cent of projects, the clients involved had been willing to invest time and effort with startlingly successful results.

- **These 90 per cent of projects made a profit, with an average sales increase of 41 per cent.**
- **The average payback period was 15 months from launch.**
- **48 per cent of projects recovered their total costs within a year or less.**
- **The average cost of a successful project was about £60,000; the average cost of a failed project about £8,000.**
- **25 per cent of projects opened up new home markets.**
- **10 per cent of projects led to reduced production costs.**

There were three key conclusions of the survey:

- **The financial risk involved in all types of design is low.**
- **The financial case for individual firms investing in new and improved designs is very strong.**

- **Small firms are just as likely to produce profitable projects
  as medium-sized or large ones.**

When measured against the high failure rate of innovation in general (more than 60 per cent of new ideas are thought to fail in the marketplace), design emerges as a risk *reduction* exercise.

TAKING THE INITIATIVE

The timing of a design project is most often dictated by a new idea or the need to revitalize an old one. Decisions to spend money on design are, in fact, most often taken in adversity – in response to intense competitive threat, for example, like the family store whose next-door plot was bought by John Lewis; or because of low market awareness, like the cheesemaker whose Stilton needed packaging that would lift it out of obscurity and help it to compete with the six other UK Stilton producers; or in the face of imminent collapse, like the battery charger manufacturer whose product Halfords (the automotive accessories chain) refused to stock because of its sharp edges.

As with any professional service, however, using design on an unplanned basis does not represent good value for money. Besides the obvious long-term disadvantages of losing the initiative to competitors, the costs of a sudden burst of design activity against a pressing deadline will be higher than those of a planned design programme.

Clients typically approach designers asking for a specific end product, but there may be more to be gained from inviting the designer in at the preliminary, brainstorming-the-problem stage. He or she may be able to help identify new market niches, and develop concepts to fill them; to help with issues of branding, such as devising names for new products or services; to draft design briefs (even if they may ultimately be placed with other designers, such as an in-house team); or to propose appropriate test exercises in advance of major projects. New twists on basic design expertise are springing up all the time. The consultancy Dragon International, for example, hesitates to call itself a design group even though it employs designers: one of its main interests lies in helping clients to review the environmental impact of their business activities.

'I sometimes detect that the client writes the brief and hands it over, then the designer comes back with a "This is my best shot" approach. I've never worked on that basis. I'd rather bring the designer in on the business issue and brainstorm the problem that we've got. I see designers as extensions of my own team. Some will want to stand outside and give you the solution as they see it, and if something isn't right they say "Oh, well, we didn't pick that up from your brief". I expect them to be more accountable than that.'

Roger Cooper
Ideal-Standard

Within each area of expertise, you don't have to have the 'full service'. You may want only an analysis of market opportunities, and advice on what your design strategy should be. Or you might want the physical design of something already defined.

Many businesses, especially those involved in manufacturing, believe that they shouldn't bring in a designer until they themselves have come up with a fundamentally new proposition. But new products, or refinements on existing ones, don't have to wait for such an innovation: they can often be improved in other ways.

'It is a commonly held view that new products have to be technically innovative, but very many successful new consumer products use existing manufacturing techniques and materials to excellent effect. The designer can excite the customer by improving performance and making sure that the aesthetic is appropriate.'

Ben Fether
FM Design

Most markets have room for something new. John Hawker, director of industrial design consultancy Design Technology, is surprised by the conservatism of many clients: 'It has always fascinated me why companies so often stay in the same market when they are surrounded by hundreds of opportunities. Often clients come to me and say, "This is the company, these are the resources, have you got any ideas for things we could make?"'

Many businesses fail to realize their design opportunities. Even where a company's product or service is superb, there may be an opportunity to project its excellence through better presentation.

Rather than specifying an end product, therefore, clients may find it more useful to start with their strategic goals. It is quite acceptable to ask a designer to address how your business might achieve any or all of the following:

- **increased sales**
- **new markets**
- **increased market share**
- **added value**
- **established or re-established presence in the marketplace**
- **rationalized systems**
- **cheaper/easier production**
- **safer/greener products or environments**
- **improved productivity**
- **added job satisfaction**
- **greater efficiency.**

Consider these examples, which demonstrate the value-adding potential of a flexible approach to design.

A product designer was asked by a large American company to design a set of darts as a promotional giveaway. The darts sets he came up with, designed for very cheap production, were so covetable that the client thought better of his original idea: he now sells them at $200 a time.

Benefits

- **added value**
- **new markets**
- **re-establishment of presence in the marketplace**
- **increased sales**

The Singapore Trade Development Board, which manages the world's busiest port, asked a graphic designer to rethink Singapore's import and export paperwork. The designer reduced the number of forms from over forty to six, radically simplifying procedures and stimulating trade.

Benefits

- **reduced costs**
- **re-established presence in the marketplace**
- **rationalized the system**
- **improved productivity**
- **greater efficiency**

Interior designer David Leon was asked to redesign the Unilever research laboratories at Port Sunlight. He amalgamated two departments into one occupying 30 per cent less space, while also accommodating 200 additional scientists, with no loss of space to anyone. The most senior bench scientists were brought closer to the point of research; new carpets and screens humanized what had been a semi-industrial environment. Refurbishment costs were less than average. The alternative would have been a further building at a cost of £10m.

Benefits

- **reduced costs**
- **added value**
- **greater efficiency**
- **greater productivity**
- **greater job satisfaction**

These examples are not untypical. They show some of the ways in which clients can get more benefits than they bargained for from a design exercise. Where you can, try to step back and consider your design needs 'in the round', as well as the problem with an immediate claim on your attention.

THE DESIGN AUDIT

To achieve this rounded perspective on design needs, the design audit or review is an increasingly popular preliminary to commissioning design services. A design audit helps businesses to understand the opportunities available to them through design. A thorough audit should assess your current resources and skills (in marketing, design, development and production, for example); judge whether proposed designs are likely to meet market expectations; look at whether 'time to market' is an issue and if so what needs to be done to reduce timescales; and determine what new skills, resources and funding (internal and external) are necessary to exploit design fully.

Businesses can of course review these issues for themselves. A 'design review kit' developed for just this purpose is available from the Chartered Society of Designers' Design Management Group (see Addresses). An objective eye may be better, however. Most design consultancies will undertake a design audit for a reasonable fee, as will the Design Council (see Addresses).

WHY USE OUTSIDE CONSULTANTS?

Businesses that have in-house designers would be right in thinking that they should be influential in any corporate design project. Clients can obviously expect from their in-house designers a high level of understanding of existing designs and their problems; they cannot, perhaps, expect an equally high level of understanding of their opportunities. This is not to say that in-house designers are less able than independent designers; far from it. However, in-house designers are not usually exposed to market forces outside their own organization. Added to this it may be difficult for in-house designers to command the same respect as consultants brought in from outside, and to persuade colleagues they may see every day to give them a detailed brief.

Consultants can bring in expertise not available in-house, and are usually able to operate without the constraints of internal pressures and politics. Roger Cooper, sales and marketing director for Ideal-Standard, explains his own reasons for looking outside: 'I honestly don't believe that you can keep going back to the same in-house person and saying, "That design was great, but now go back and do me a better one, and a better one, and a better one still". There has to be some fresh input.'

The main disadvantage of bringing in a design consultancy is that the client is forced to put a lot of effort into briefing. This can be so labour intensive that some clients feel they might as well do the design themselves (and often do, unfortunately). On the other hand, being forced to put a lot of effort into writing a brief is never a bad thing.

Usually, where there are in-house designers, they will work as a team with the external consultancy. This situation may need careful handling. The decision to bring in an external designer can be a sensitive one, perceived as a slight upon in-house expertise. This can escalate if the external consultancy goes on to pursue press coverage which suggests the project was all its own work.

2 finding your designer

2 THE DESIGN CONSULTANCY MARKETPLACE

Britain is exceptionally well supplied with design expertise, having over 3,000 design consultancies, more than 40,000 trained designers, and 7,000 students graduating into the profession from British colleges every year. In all disciplines the reputation of British designers is world class, and many now find much of their work outside the UK. Some consultancies have international networks of offices. UK-based groups, and particularly graphic design consultancies who form the majority, are most concentrated in London and the south east of England. Scotland and Wales proportionately have fewer consultancies and among them a higher concentration of product designers.

The average trained designer will have completed a three- or four-year course, probably including some work experience, usually resulting in a diploma or a degree. Within each of the broad areas of product, graphic and interior design are many specialisms, and once out in the world, practising designers are likely to specialize still further. Most designers are competent at tackling a broad range of projects within their general discipline, and some may be happy to tackle completely new territory. Although rare, it's not unknown for car designers to design furniture, and for graphic designers to design products, for example. The trend is, however, towards greater specialization.

The average length of time a British design consultancy has been in operation is 12 years, making it a young business. There are consultancies which have been established for 50 years and more, but their rates tend to be high. Design buyers new into the field are more likely to be dealing with less experienced, and therefore cheaper consultancies, where standards are more variable and client input more critical.

The average design consultancy probably employs less than a dozen people and many consultancies are solo operations. The bigger the consultancy, the more likely it is to have a formal structure, which could include strategic advisers, account handlers, business planners, researchers or project managers, any of whom may have some intermediary role between client and designer. Inevitably this scale of business means bigger overheads and bigger fees. Some of the largest groups are public limited companies, though this is not necessarily an index of success.

SOURCES

The most popular methods of finding designers are through:

- **word of mouth**
- **an advertising/PR agency**
- **a design organization/institution**
- **a direct approach from a design consultancy.**

Of these, the advertising/PR agency route is thought by many to be the least successful. One PR group has claimed (though this is disputed by designers) that PR companies are the brokers in 98 per cent of graphic design projects. The agencies often have no specific expertise in design broking and have the reputation of choosing only those consultancies with whom they have connections. PR and advertising groups are also often accused of inviting presentations from more consultancies than necessary, or of angling for the work themselves; in the case of exhibition design, packaging design and corporate identity, there may well be a conflict of interest.

Details of the following sources are listed in the Publications section on page 78 and the Addresses section on page 85.

Design brokers

The Design Council

The Design Council, founded in 1944, exists to promote the effective use of design in industry and to foster design education. The Design Council has a virtual monopoly on design brokerage in the shape of its Designer Selection Service. It maintains the largest available register of consultancies (about 850), whose output the Council has vetted for design capability and efficiency of working methods. Consultancies pay to be listed on this register. Clients' requirements are matched against the experience of the consultancies on the database and, for a fee (£80 in 1992), clients are given a shortlist of three consultants whom the Design Council considers appropriate for the project. The service also includes an element of design management advice; staff will assess a client's needs, attend consultancy presentations and may arbitrate in the event of disputes.

The Design Business Association

The Design Business Association (DBA), formed in 1986, represents the UK's design consultancy sector. The DBA has a membership of over 200 design consultancies and holds a database of 300 other design businesses. While it has no formal recommendation mechanism, the DBA publishes a directory of its members which is free to clients, and has produced a series of guides to choosing and working with design consultancies.

The Royal Society of Arts

The Royal Society of Arts (RSA) has a broad role in encouraging the arts, manufacturing and commerce. The RSA runs a student awards scheme which places winning student designers on 12-week attachments with direct companies. Details are available from the RSA.

Publications

The magazines *Design* and *DesignWeek* publish league tables of British design consultancies. *Design* magazine also publishes a guide to design consultancies in Europe three times a year. There are several directories of creative services suppliers which include design consultants.

Databases

The Applied Art and Design Index has on its database over 200 art and design publications internationally, and will run a search (if you wanted to find out what had been written about a given consultancy, for example) for free.

Creative Information Databases runs a compact-disc-based reference system which holds details of the work of designers, photographers and illustrators. Samples of work are viewable on the computer screen.

Regional design centres

There are several regional design centres which are focal points for local design activity, and which house clusters of design consultancies; for example the Bradford Design Exchange; and DesignWorks in Felling near Newcastle. Local chambers of commerce should be able to put prospective clients in touch with their nearest centre. Such centres are not able to recommend specific consultancies but may nevertheless be worth approaching for contact names and addresses.

Events and exhibitions

Student shows

Some clients approach students for their design work, either on a freelance basis, as a fully commercial project in collaboration with the college, or as a result of holding a student competition. (Anita Roddick reputedly paid £25 for the Body Shop logo, designed by a student of Brighton Polytechnic many years ago.) College tutors and students welcome involvement with 'real' projects, but there can be drawbacks for the client. There is now an abundance of student design competitions and you should not expect a huge response to yours if you mount one. You should also accept that expectations of technical perfection and easy and cost-effective production have to be tempered when commissioning inexperienced student designers.

These reservations aside, showcases for student design work are worth visiting to see new approaches to design. Design college degree shows take place every year in the summer months, with many regional colleges mounting shows in London as well as locally. In May each year *Design* magazine publishes a list of all such shows throughout the UK.

There are three main venues which seek to present a compilation of the best new student work each year, from colleges around the UK: the Design Council's Young Designers Centre; the independent New Designers Exhibition which takes places every July; and on a smaller scale, furniture retailer Zeev Aram's annual selection of the products and furniture he considers interesting. All three venues are in London.

Professional exhibitions

The Design Council mounts temporary exhibitions in London, usually on themes related to the manufacturing industry. Selected exhibitions also tour outside London.

The Design Museum, founded in 1989, exhibits permanent and changing collections of both contemporary and historic mass-produced designs from around the world.

Consultancy initiatives

Consultancies are becoming more adept at getting their names known through 'cultural' routes – hosting discussions and open days, sponsoring exhibitions and publications – and these may be a useful introduction to their services.

Press coverage

The specialist design press and the national press are useful sources of contacts and information. *The Financial Times* publishes a design supplement annually in October and sometimes features design projects in its management section; *The Times* covers design on Tuesdays; *The Independent* and *The Guardian* both also give occasional coverage of design.

CHOOSING BETWEEN CONSULTANCIES

'I go looking for design firms that are on heat.'

Jane Priestman
Design Management Consultant

Even large, experienced companies can make mistakes in their choice of consultant. One international pharmaceuticals group has had little return on the £200,000 it spent, over a period of two years, on the services of a renowned design consultancy; one of the privatized water authorities has had a similar experience. The likelihood is not that these consultants gave poor service, but that they were not well matched with the client organization. It pays to take great care in your selection of consultancy. The first step, having identified perhaps six groups whom you think may be appropriate, is to request a brochure from each, or glean information over the telephone. From the initial six you should shortlist up to three, who should be comparable in size and specialism as far as possible. There is little point in shortlisting a one-person operation, a medium-sized interior design practice and a large firm of architects, for example. You should then invite the three consultants to give a 'credentials presentation'.

The consultancies will normally travel to your office and will each need roughly an hour of your time; it would be difficult for them to do themselves justice in less. Credentials presentations, which are free, nevertheless involve some planning on the part of the consultancies and you too should be prepared to answer questions on the project you have in mind.

Most fundamentally, the consultancies will want to be assured that you have the money, the authority and the need for their services. Be ready, therefore, to discuss the following points.

Budget

What is the total expected investment in the project? Is it possible to say how you expect this sum to be divided between design costs and implementation costs? Money may be hard to find if you are calling in a designer as a response to an unforeseen problem, but you must give some idea of what is available. Will price be the main criterion upon which you base your buying decision? And if price isn't the key factor, what is?

Authority

Will you be in sole charge of the project? Design projects often founder when the person who has to approve the final design is not the same person who gave the brief. You should explain your role and who else may be involved. If others are involved, aim to set up a small panel of relevant people to take part in the briefing process. You should certainly try to involve the people who will produce, distribute and promote the design.

Objectives

What is the purpose of the exercise you have in mind? How can success be measured? Give the consultancies some background on your business and the challenges it faces. Explain how you see your company – its way of working, its 'personality', its aims and resources – and where the proposed design project fits in.

Previous or current use of design

What arrangements do you have for design services at the moment, if any? How would you describe your corporate identity? (As designers are fond of pointing out, all companies have a corporate identity whether they think they have one or not – an identity made up of everything from how the telephone is answered to the type of flooring in the reception area). Perhaps some corporate identity guidelines already exist in published form, maybe as a result of a previous design exercise. If so, what elements of the identity must be retained?

Schedule

Outline your proposed timescale, mentioning any interim deadlines and any plans for a formal launch of the finished design.

GROUND YOU SHOULD COVER WITH DESIGNERS

What follows is a list of the factors you may want to take into account when you make your choice. You can cover this ground informally; alternatively, you may want to run through it question by question, using it as a checklist at the presentation. Or you may even want to edit it into an appropriate questionnaire, to be sent out to consultancies before you invite them in. Peter Clapp, design manager at W H Smith, finds this a useful first step in selecting some of the groups he works with.

Profile of the consultancy

Establish the size and structure of the group: how many staff it has and how many of these are principals (directors/partners), design and technical staff, and administrators (account handlers or equivalent). Find out who would actually be assigned to your project and ask for an outline of their qualifications and experience. Enquire about professional affiliations – designers may, for example, be Design Council registered, DTI Initiative registered, members of a chamber of commerce, members of the Chartered Society of Designers or Design Business Association, Fellows of the Royal Society of Arts, or (most prestigious of all, and few in number) Royal Designers for Industry.

Fees/charges

Ask the hourly rates charged for directors/partners, administrators, and designers – different rates may apply for senior designers, middle rank designers, and junior designers/trainees. The consultancy will probably charge a daily or hourly rate which is an average of all of these (see Costs, page 29), but it is helpful nevertheless to understand the charging structure.

Working practices

Does the consultancy encourage direct contact between clients and designers, or do communications tend to be channelled through a third party (such as an account handler)? Most clients and designers seem to find that direct contact produces the most favourable results.

Establish how the consultancy likes to work: is there a formal design review process which ensures a progress evaluation, and if so how frequently are such reviews held? What sort of project documentation is kept? Can the consultancy show evidence of implementation skills on long, complex projects?

Technology

Many consultancies have computer aided design facilities of some description. You will almost certainly find graphic designers working on Apple Macintosh computers, which have become the industry standard.

Although many first rough ideas are still sketched with a marker pen or hewn out of plastic foam, designers increasingly use computers to develop their ideas into realistic form and to try out variants in a fast and cost-effective way. For the client, the use of computers in design studios offers several benefits.

The first is visualization: you can look at your design taking shape on screen and give the designer instant feedback. The immense flexibility offered by some computer aided design programs means that forms, colours, reflections, shadows and textures can be shown in fine detail and adjusted at will; some CAD packages can accurately show not only exterior finishes but internal geometries and movement. Industrial designers may hold on computer a library of materials so that different finishes can be tried out on screen.

For textile designers, CAD offers the possibility of scanning in existing designs and manipulating them to show how the cloth will fall when wrapped around a piece of furniture, hung in an interior or draped on a model.

The clients of interior designers may be able to see an on-screen 'flythrough': that is, they can view a proposed interior from every perspective (with appropriate lighting effects) as if actually walking through the building. In all areas, the visualization potential of computers is developing very rapidly.

Computers also have a major role to play in cutting lead times. As well as saving time during the design process by eliminating much checking and modification of drawings, time can also be saved on feeding specifications into the production process. Data for graphic images or packaging projects can be output directly to video, film or high resolution print; data for product specifications can be output directly to computer numerically controlled machine tools. Effective use of CAD by a consultancy can decrease product development time by several weeks.

It should be said, however, that CAD systems vary greatly in their level of sophistication and the leading edge systems are beyond the financial reach of many design consultancies.

It also pays to quizz consultants on their use of the technology they do have – often the presence of CAD facilities is used as a selling tool, but is underexploited in practice.

Compatibility

Very often it is personal rapport which prompts a client to hire a designer, especially when choosing between equally competent consultancies. Getting on well together is important; you are looking for a business partner. As many design projects develop, the designer will become very aware of the client's inadequacies, in terms of indecisiveness, lack of clarity of objectives, and general disorganization. There will be tough times in the course of any design exercise, no matter how well-managed the client company. When things go wrong, will the designer respond positively to problems and work well with the management team?

'At the end of the day I feel the consultancy has got to be easy to work with. In my experience, when you start out on a project, you never know exactly what you want or what is possible. Consequently, designers will become familiar with the client's predeliction for changing the goalposts. Some people are emotionally better equipped to deal with that than others. One of the reasons I like dealing with our present consultancy is that not only are they very good designers but it's almost impossible to upset them.'

Steve Hallam
Fisons

Degree of specialization

Check that design is the main activity of the business you approach. Many businesses with some involvement in design may offer the service themselves. Printers, for example, often offer corporate identity design.

- **Do you need experience of a particular market or a particular country?**
- **Is internationalism important?**
- **In which sectors is the consultancy most experienced?**
- **What in-house services (for example research) does it offer?**

What proportion of the group's work is in:

- **Corporate publishing?**
- **Direct mail?**
- **Exhibition/display/point of sale?**
- **Packaging?**
- **Retail interiors?**
- **Office interiors?**
- **Consumer/contract products?**
- **Industrial products?**

(This list can be refined as you see fit.)

If it seems that the consultancy you're talking to clearly lacks the specialist skills you're looking for, ask where they would go in your position. Reputable design groups will be straightforward about what they can and can't do and therefore should not be offended by this question.

If you need to engage more than one designer, you have the option of choosing them yourself or getting one of your chosen consultancies to put together a team. Design consultancies are getting increasingly used to working in project teams in this way.

Recommendations

Talking to a consultancy's existing or former clients is a valuable exercise. Ask the consultancy for the details of several former clients whom you can approach for information. At the least, you need to know that your chosen consultancy is financially and organizationally sound.

Profile

Client companies who can afford it very often choose a well-known consultancy to bolster confidence in the project. A high profile doesn't guarantee better design skills, however; there are many smaller groups as talented as the big names.

Design awards

You might want to consider approaching consultancies on the basis of their success in award schemes. Don't, however, disregard consultancies that don't hold awards. Some consultancies have a policy of not entering for them; some awards reward the fashionable rather than the durable. Among the many national and international design award schemes are the British Design Awards, run by the Design Council, which recognize a broad spectrum of achievements spanning function, appearance, quality and safety; and the Design Effectiveness Awards, run by the Design Business Association, which assess design projects on the basis of their commercial effectiveness.

Involvement in strategy

To an extent, all designers get involved in strategic decisions, particularly those relating to marketing. The analysis phase of any design project throws up difficult questions, the answers to which may have a significant impact on the client company. This is most visibly the case with corporate identity projects, in which the designer has the specific remit to analyse corporate reputation and how well it reflects corporate objectives. Product designers too can influence company strategy to an extent not often anticipated by clients (for example, by developing from one core product a series of variants suitable for new markets).

Some design consultancies (usually the corporate identity specialists) relish this aspect of their work, make it very much part of their offer and charge for it accordingly. Other consultancies (usually the product design consultancies) may become involved with the strategy but are less skilful at packaging it as part of the whole service.

'We often undershoot on our quotes because our fees haven't kept pace with the position we adopt within the company. We get very

involved with strategy, and that's where a lot of our money is going because we don't know how to build that into the quote. The more we try to build it in – the more we try to reflect what we actually do – the more we are at the mercy of consultancies who are cheaper because they don't get involved with strategy.'

David Edgerley, Isis

Some consultancies, by contrast, are reluctant to interfere directly in a client's business plans.

'We have had some very big clients coming to us on the rebound, saying effectively, "Do the design, but don't tell us how to run our business". We as designers are in the business of making other businesses successful, and so naturally I am interested in the sales performance and the thinking which drives the company. Companies are pleased that I'm interested, but they know where my real strength is. Once designers start getting involved with this, in no time at all they start to think "Hey, I'm a businessman". A lot of designers will be respected and treated as members of the management team in the client company, but to posture as a business adviser is dangerous, and I would not counsel it.'

Ken Grange
Pentagram

In general, many more design consultancies now position themselves as strategic advisers than would have done five years ago. The aspiration is probably a good one, but in practice there are some humble operations laying claim to skills which suggest that their directors should be running ICI. Let the buyer beware.

Location

Most clients use local design resources, for easy access and communication. But choosing a designer in another part of the country, or even in another country, need not present problems. A great deal of design work is conducted by fax. John Benson, assistant director, the Design Council's industry division, puts the order of desirability thus: 'First, good and local. Second, good and not local. Worst, local and not good.'

Premises

You may want to, and probably should, visit the offices of design consultancies you are considering. This is standard practice and what you don't want to find is a rented garret occupied by temporary staff.

Hands-on experience

Some product design consultancies have undertaken manufacturing projects themselves while some interior design consultancies have opened their own shops. These designers turned entrepreneurs claim a more informed understanding of the client's perspective and a keener grasp of issues such as pricing and distribution.

Creativity

> *'For me personally, my ability to judge on the basis of a portfolio is limited. Everyone has a decent portfolio; they're not going to show you any rubbish. I tend not to pay much attention to them, although some of my colleagues do look at them carefully.'*
>
> **Steve Hallam**
> **Fisons**

Creativity is consistently judged by clients to be the most important attribute of a design consultancy (more important even than fees) yet is not always easy for inexperienced clients to recognize. Many people are dazzled by portfolios; they all look so good that clients feel reluctant to ask questions. Clients should look for originality; lateral thinking; strong problem-solving ability; conceptual range (there shouldn't be too much of a recognizable house style); high-quality execution; good production and high standards of presentation. Ideally, the designer should focus on two or three projects only and talk the client through these in detail, quantifying their effectiveness.

Occasionally, where designers feel that their portfolio of implemented projects is not truly representative of their talents (because clients have watered down their design concepts), they may include some work produced speculatively to a self-determined brief.

At the other extreme, there is a trend in graphic design towards designers dispensing with the portfolio at presentations in order to put themselves on a business level with their clients, but clients usually expect to see some past creative work and designers usually oblige.

PROPOSALS

The next stage, after the consultancies have gone away with a clear idea of the project you have in mind, is to invite written proposals from those you favour. Making bids for work represents one of the design consultant's major costs, and this should not be abused. Many consultancies prepare proposals in minute detail and devote substantial resources to creating an appropriate document – often running to several thousand pounds in staff time devoted to research and writing. Consultancies will not expect to be asked for actual design work at this stage, unless you commission it on a paid basis. You may tell the consultancies how many groups are in the running but should not name names.

Proposals will normally be presented to you by one or more members of each group. These meetings vary in format, but should always allow a dialogue between the client and the consultancy. Each consultancy should summarize the issues faced by you and your competitors; explain its perceptions of your company and its design objectives; and give an outline of the proposed design work and how the client-designer relationship will work in practice throughout the project.

All proposals should set out the following in writing.

The project objectives

Project objectives should ensure that both parties understand and agree the ultimate aims of the project. Graphic design projects in particular frequently proceed, disastrously, on the basis of an unrecorded discussion. It is important to define the scope of the work, and to agree on how fundamental is the designer's remit. Is it a design-from-scratch project, or a more cosmetic redesign?

The programme of work

The programme of work will cover the work to be done, when it will be handed over and in what form. It should detail the required standard of work, including the stages along the way: will rough drawings do, or do you need a lot of detail? It is helpful to have specified in the programme of work the stages that will *not* be included.

The schedule

The proposal should set down the number of consultancy days which the design project is expected to occupy, plus a start date and completion date.

Fees and other costs

The basis on which fees and other costs are to be calculated must be included, along with the stages at which payments are to be made. Most design charges are levied in instalments related to the progress of the work.

Some fees quoted may apply for a limited time only and if the project is delayed costs will have to be renegotiated. If the brief changes, the consultancy should requote.

Some advance payment may be stipulated, especially if you and the consultant have not worked together before.

Ownership of the work

It is vitally important to establish who will own copyright, or other rights, at what point and for how long. Commissioning design work does not automatically give you full rights over its use, unless negotiated and agreed in writing by the designer. The client may only have limited rights, for example, to use the design in question for a limited length of time, in a limited geographical area, or for a particular purpose. If clients transgress these limits they may be sued. If the designer wants to restrict rights, the client should negotiate a lower fee to reflect this. Make it clear at the outset what rights you expect (see Legal Issues on page 60) and establish whether the designer will grant them. Normally consultancies are willing to assign full rights of the work to their clients, but may request a fee in addition to those already mentioned.

Confidentiality

Designers should disclose any involvement with rival or related companies, suppliers or specialists. Designers should agree to keep confidential all information relating to clients.

Terms and conditions

The proposal should be accompanied by the consultancy's usual terms and conditions. Most consultancies will have ready-prepared standard terms and conditions which they ask all clients to sign. Typically, they cover methods of payment and interest charged on late payment; notice periods by which the client must communicate any dissatisfaction with the work delivered by the consultancy; and notice periods which must be given to the consultancy if the client abandons the project.

There is usually some negotiation on both the proposal and the terms and conditions. You may want to assert your own terms, for example. Clients' terms and designers' terms cannot both apply, however, so in this situation you could negotiate a version including elements of both.

Agreement, once reached, should be put in writing and signed by both parties. This is the contract between you, and it can then only be changed by mutual agreement. Standard terms cannot be imposed once a contract is made.

Make it clear who in your company can and cannot instruct your designer: a contract can legally be established verbally, in person or by phone. It is unfortunately the case that many design projects get under way without any form of written agreement. This is partly the fault of design consultancies who are lax about the written proposal and partly the fault of clients who don't want to be bothered with drawing up a contract, or who draw one up only to ignore it later.

FREE PITCHING

There has been a long-running debate over the practice of creative free pitching, whereby a client invites several consultancies to produce finished design solutions in competition, and on an unpaid basis, for a job. Free pitching is routine in the advertising industry, and many clients expect design consultancies to offer the same service.

Advertising agencies ultimately charge higher fees than design groups, and so are able to subsidize their free pitching. Design consultancies that invest heavily in free pitching make themselves very vulnerable. Professional design associations object to the practice and are doing all they can to stamp it out, on the grounds that it devalues design in the eyes of clients and undermines competitive selection. What they mean by this is that (a) clients may get the false impression that their problems can be solved for free after a rudimentary briefing, and (b) some consultancies will pitch for free and some will not. Comparisons are made with other professions: would clients ask accountancy firms to produce a cashflow forecast for nothing?

Free creative pitches are understandably tempting for clients, particularly those who are not sure how to assess a portfolio. Within the design profession itself, opinions are mixed. Most consultancies will not pitch for free simply because they can't justify the investment of time and money in an exercise which may produce no return. Consultancies have made exceptions, however, for large and lucrative projects, or where they believed they would otherwise stand no chance of winning the work.

Attitudes are hardening, though, as a result of some particularly heinous cases, like that of three graphic design consultancies each of which invested £25,000 on free pitches, against the promise of a slice of a £300,000 corporate identity job. The client was a large building society. In the event the client chose none of the consultancies, and the matter went to court.

If you do ever ask for a free pitch, remember that copyright in any work the designers produce will remain with the designers.

PAID PITCHING

You may want to set a test problem on a paid basis, and see how it is solved by the chosen consultancy (or consultancies). This is sometimes the only way to be sure of technical competence, and may be worthwhile on a major project. This exercise will show how well the brief is understood, how creatively it is interpreted, how frequently the designer likes to communicate with you and by what method, and how professional is the outcome. Consultancies usually like working this way because it demonstrates that clients are serious about the work. And, of course, it gives you the option to back out if the results aren't up to scratch.

'We placed a test brief with two London consultancies. We wanted to find out how they would approach it. We had interviewed seven consultancies and narrowed it down to those two. It only cost us £2,000 each. But we were not impressed with the results, and decided to use neither of them. What we found was that anything they'd done which was innovative was not practical, and anything they'd done which was practical was not attractive, or else was virtually a copy of one of our competitors' products. Funnily enough they both chose the same competitive product to copy!'

Bill McMichael
Xpelair

Once your decision is made, let the rejected consultancies know that they have not been chosen. If you can also tell them why they were rejected, they will appreciate it.

3
costs

CALCULATING THE COST

Both manufacturing and service companies employing fewer than 500 people are eligible (until 1993) to apply for financial help with design consultancy through the Department of Trade and Industry's Enterprise Initiative (see Addresses, page 85).

The cost of using a design consultancy is made up of three basic elements:

- **fees for the design work itself**
- **expenses incurred in the process, including any bought-in services**
- **the cost of producing whatever is designed.**

DESIGN FEES

While design fees vary tremendously according to the size of the consultancy and its overheads, the type of client, the type of job and the discipline involved, a crude generalization might designate anything around £5,000 a small project, anything under £35,000 a medium-sized project, and anything over £60,000 a large project. At one end of the scale, for example, a survey by PR and marketing consultancy Point to Point in 1991 found that 53 per cent of its sample of companies expected to pay between £1,000 and £3,000 for a first concept pack design presentation on a single brand; 27 per cent expected to pay under £1,000. At the other, a survey conducted by National Power, *Industry and Design 1991: The Senior Managers' Viewpoint*, found that 55 per cent of its sample of UK manufacturing companies spent between two and five per cent of sales turnover on product design and development – a fairly substantial commitment.

Designers generally charge on the basis of a fixed daily or hourly rate. A typical day of 7.5 hours would be charged at between £200 and £500 (1992 prices). There are some extreme deviations from this, however. Hourly rates are on average about a third higher in London than elsewhere. The rates of the most prestigious London consultancies can go above £1,000 a day, while a one-person business in a rural area may charge less than £100.

Graphic designers charge the highest hourly rates (there are graphic designers in London charging more than £250 an hour, though these are few) with interior designers' rates running close behind. Product design rates are much lower. Many people commissioning product design are surprised to find themselves paying more for the design of the packaging than for the design of the product itself. This is partly accounted for by the fact that people commissioning graphic design are often working in a marketing environment where high spending is customary on advertising and print production; they have developed a tolerance, and often an expectation, of high fees. People commissioning product design, conversely, are accustomed to much lower expenditure on services. They are also looking at longer timescales, and looking at them in the context of high production costs (on tooling, for example).

EXPENSES

All design projects involve expenses; from photocopying and travelling to the client's office, to the cost of hiring specialist services. These are some of the services which might be called for in the course of design work.

Graphic design projects

- **photography**
- **illustration**
- **copywriting**
- **typesetting**
- **artwork**
- **origination and printing**
- **market research.**

Product design projects

- **ergonomics**
- **modelmaking**

- **engineering**
- **materials technology**
- **market research.**

Interior design projects

- **signage**
- **shopfitting**
- **architectural services**
- **furniture**
- **lighting**
- **heating, ventilation and air-conditioning**
- **market research.**

Sometimes clients pay such specialists direct. Sometimes the design consultancy will arrange payment and will usually add a handling charge of at least 15 per cent for doing so. There is scope here for large mark-ups, so if a consultancy proposes using bought-in services at any time be sure that you understand and agree the cost basis.

IMPLEMENTATION COSTS

Implementation costs can only be assessed once the scope of the project is fully defined. It is then worth getting quotes as soon as possible. (Linda Liddell, managing director of a small consulting firm, was delighted to pay £4,000 for a new letterhead and business card, but considerably less delighted to discover that each piece of letterheaded stationery cost more than £1 to print.)

METHODS OF PAYMENT

The choice of method of payment can have a significant effect on the final bill. These are the common options.

Fixed fee

A fixed fee is a popular arrangement, although it's often difficult for designers to project sufficiently far ahead to quote accurately. In these cases a fixed fee is established for the first, foreseeable stage, with hourly or daily rates applying thereafter.

Advantage
The buyer knows exactly what is being spent, and the designer knows exactly what is available.

Disadvantages
Designers may set too low a fee. This becomes a disadvantage to the client if the money runs out and the project is unfinished. While designers are still bound to complete the work, their commitment may flag. It is also sometimes argued that a fixed fee encourages designers to do the job and step back from it, rather than encouraging them to feel a continuing sense of partnership with the client.

Hourly or daily rate
These rates are sometimes used in combination with a fixed fee.

Advantage
The buyer can be sure that corners will not be cut. Time-based fees are calculated by time sheets and the buyer can see exactly what time has been spent on the project.

Disadvantages
The buyer may worry that the clock is ticking, and feel tempted to use the consultancy's time as little as possible. Also there may be the fear that a consultancy is spinning out the project unnecessarily. To avoid misunderstandings, clients and designers can agree the minimum and maximum payable sums.

Percentage
Paying a percentage is most common among interior designers and exhibition designers, who often charge a fixed fee for initial stages and ask for a percentage of the implementation costs thereafter. The percentage varies; it will be higher on small projects than on large ones.

Identity

CJS Plants is a small company which organizes interior plant displays. Its owner, Charles Short, commissioned graphic design consultancy The Partners to develop an identity that would help his company to stand out from the crowd. The design group provided a witty solution using green fingerprints on stationery and clustered around the handles of the delivery van; and the corporate brochure features plant drawings on textured paper and is bound with a bamboo spine, continuing the garden theme.

Says Short: 'Investing in something like this is a big step for any small company, but it's been brilliant; people remember it. It has had a direct effect on turnover. The identity is a continuing thing and the design group has more ideas, which are wonderful, in the pipeline.'

Advantage

The amount to be earmarked should be reasonably clear from the start if rough specifications for the work are known.

Disadvantage

The amount paid may bear little relation to the amount of time actually put in by the designers.

Royalty

Royalties usually only apply in the case of consumer product design. There are consultancies that work primarily on a royalty basis, and clients may sometimes find that such consultancies want a significant amount of control over the design project. Usually the agreed amount is a percentage of the net ex-factory price, and is in the range of 1-5 per cent. Royalties will be lower on high-volume products; higher on short runs.

Advantages

The prospect of royalty payments may motivate the designer to design well and to design for the long term. For the client there is no fee to pay upfront. A royalty arrangement can foster long-term, mutually supportive relationships between clients and designers.

Disadvantages

Some designers feel that royalty agreements push the risk onto them without the prospect of a significant share of the rewards. For the designer, royalties also mean very long periods of pay-back, which may present cashflow problems too severe for a small consultancy to handle. The success of the arrangement will depend on how effectively the client company markets the product, but because the client doesn't have to pay anything upfront, the arrangement may in fact lessen the client's feeling of responsibility towards the project.

Consultancy fee or retainer

A consultancy fee or retainer does not usually cover actual design work, only advice, and is paid annually or quarterly.

Keeping a designer on a retainer happens most often when a client has worked with the designer before and wants to be able to refer general questions of design strategy. Clients may also see the retainer as a means of preventing the designer from working on projects for a rival company.

Advantage

Retainers promote a continuing sense of partnership between designer and client.

Disadvantage

There is potential for clients to abuse the relationship in their efforts to get value for money.

Quotes and estimates

Designers base their quotes on precedent, or if there is no precedent, on an educated guess at how long the work will take. Quotes may also depend on other factors, such as whether or not the project is a one-off. Some consultancies may deliberately charge more if there is no follow-up work.

Some clients are notorious for playing consultancies off against one another: for example by getting quotes from small unknown consultancies (whom the client has no intention of employing) and using those quotes to beat down the major names. It is worth remembering that anyone who decides to offer a service at less than its market value is unlikely to give their best. Some consultancies will in any case withdraw from the fray if they suspect a client is 'playing the market'.

If keeping costs down is an overriding priority then a small, young consultancy outside London may be the obvious choice, though always be aware that inexperience may have time-consuming consequences.

It is not unusual for design buyers to receive different quotes in response to the same brief or specification, as the design manager of a Midlands manufacturing company found out.

'We sent out an identical brief to two design consultancies. In response, one consultancy quoted £98,000, the other, £14,000. And

Packaging

Multi-disciplinary design consultancy Minale Tattersfield has given Irn Bru, Scotland's favourite soft drink, its first new identity (opposite) since 1949. The aim was to keep existing customers happy, but expand the appeal of the drink throughout the UK and continental Europe: it therefore had to look like a major new brand while keeping its traditional brand values. Among the steps taken to achieve this, 'unsophisticated' imagery was removed and the human figure, originally derived from a Victorian engraving of a weightlifter, became an 'athletic, asexual' character felt to be more appropriate to today. Sales of Irn Bru are growing at an unprecedented rate of 76% per annum.

BARR
ORIGINAL AND BEST
IRN-BRU
SERVE
COOL
FLAVOURED · SOFT · DRINK

the more expensive one wanted us to do most of the work ourselves! In my opinion neither of them had understood the project properly and what was involved. I estimated that the true cost should have been about £40 – £45,000.'

This client was further perplexed by the fact that the low quote came from a London consultancy, while the higher one came from a local group.

In cases like this, ask the consultancies involved to itemize their quotes. Wide variations are sometimes explained by the fact that one group is including all possible elements of payment, while another quotes only a fee based on its hourly rate. Sometimes discrepancies or omissions in the client's brief are responsible for the differences in quotes. It is common for clients to proceed from complete 'design illiteracy', while talking to the first consultancy they meet, to relative 'design awareness' by the time they meet the third. This may cause clients, without realizing it, to issue briefs of varying degrees of refinement. For this reason, some consultancies are less than keen to be first in line.

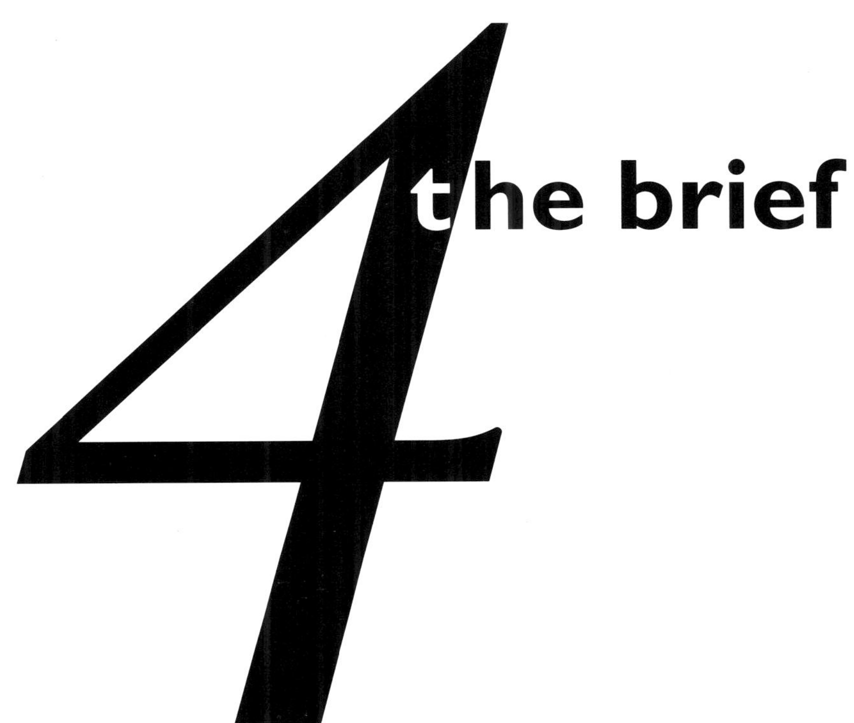

4 the brief

4 BRIEFING CHECKLIST

Considering the following points should help you to prepare for the briefing process. Not every point will be relevant to your project, and some questions will not be answerable at early stages, but it helps to be aware of them. Many of these issues are interrelated.

YOUR MARKET

What does the end user want?
What are your competitors doing?
How do you want your design to be perceived in comparison with theirs?
How will the special properties of your design be communicated in the marketplace?

YOUR AIM

What do you want to achieve? How can success be measured?

YOUR BUSINESS

Must the design conform to: a) existing corporate identity guidelines? and b) an existing sub-brand?
Will it be: a) a one-off? or b) part of an existing series or range?
Does it have potential as a starting point for a new series or range?
Could common elements produce economies of scale?

YOUR PROPOSED DESIGN

Functions/performance standards
Safety/efficiency
Context
Problems/constraints
Brand values
Format
Number to be produced
Adaptability
Special features/options
Size/weight

SCHEDULE

Start date
Research schedule
Critical approval phases
Delivery date
Pre-launch deadlines (eg
presentation to colleagues)
Market launch
Lifespan
Evaluation
Start date of successor design
(industrial design projects may
proceed simultaneously so
that products 'leapfrog' one
another)

BUDGET

Research fees
Design fees
Legal fees
Bought-in services (eg model-
making)
Pre-production processes (eg
origination for print, tooling
for manufacture)
Production/implementation
costs
Testing
Launch
Evaluation

LEGAL ISSUES

Contractual agreements
(between client and designer)
Clearance to proceed (eg
planning consent)
Establishment of rights (eg
patentability, copyright
assignment)
Legal requirements (eg health
and safety legislation)

SENSITIVITY

Is the design sensitive to its
users, both physically (eg easy
to understand and use even
for people with less than
perfect sight or dexterity) and
culturally (respectful of
international, national or local
custom)?
Is the design sensitive to the
environment, in the course of
its production, in use, and
after disposal?

PRODUCTS

Ergonomics
Mechanics/electronics
Maintenance/cleaning
Styling
Materials
Colours
Finishes
Product graphics/labelling
Packaging
Instructions
Transportation
Merchandising/point of sale materials

PRINT

Key messages
Amount and style of written material
Legibility
Relationship between text and images
Number of languages
Charts/diagrams
Illustrations/photographs
Folding
Binding
Materials
Colours
Finishes
Postage/transportation
Display

INTERIORS

Circulation
Access
Heating/ventilation/air conditioning
Information technology/cabling
Lighting
Entrances, exits, fascias, reception
areas
Materials
Colours
Finishes
Furniture
Sight lines
Signage
Lifts
Toilets
Acoustics
Maintenance
Cleaning

POINTS TO CONSIDER

'I always thought of design as an isolated technical exercise focusing on the production of technical drawings. But I took away the idea that design is an overarching concept which can control a company's identity on all fronts: publicity, environment, technology.'

**Participant, Chartered Society of Designers'
Design Management Seminar 1991**

More problems are caused by inadequate briefing than by anything else in the design process. Even when a written brief does exist – which is by no means always – the language in which it is expressed can be a problem. Take the word model: does this mean a small-scale model, to give an idea of the concept? A full-size model showing the detailing? A working prototype? It is common for client and designer to have different expectations without realizing it until the project nears completion. So, however you choose to communicate your brief, be specific.

The best place to start when drawing up a brief is with the end user. It is perhaps surprising, but many companies have no clear idea of who they are selling to.

There are four main types of research which may be relevant to the planning of a design project.

- **General business research – economic forecasts and industry snapshots.**

- **Market research – establishing the size of a particular market, the trends within it, the key players and their market share.**

- **User research – assessing how an existing product or service is used and regarded in the marketplace.**

- **New product research – inviting specific comment on a proposed new product (or service), its pricing, packaging, standing in relation to competitive products, and completeness of the range.**

The Market Research Society and the International Market Research Association hold lists of professional market research organizations. The daily rate for research is roughly comparable to that of design consultancy and the principles of selection are also much the same – initial exploratory conversations are free, but you should be prepared with an outline brief. If you do commission research, ask for the findings to culminate in a personal presentation as well as a written report.

Some design consultancies have affiliations with market research groups and may be able to obtain their services at preferential rates. Other design groups may have in-house market research capability: one well-known group employs a psychologist. Most of the market research and marketing advice available from design consultancies tends to be of the intuitive kind, obtained through informal, qualitative interviews, rather than the type that stems from head-counting.

There are also research organizations which specialize in areas particularly relevant to design, for example, forecasting colour and style trends (see Addresses, page 85).

Sometimes it will be marketing managers, experienced in commissioning market research, who handle the commissioning of designers. The art here is in reconciling the known facts established by research with the as-yet-untried creative solutions of designers. Market research managers tend to be notorious for their faith in statistics; designers notorious for their conviction that their instinct is a better guide.

> *'Our biggest threat was ourselves. Our marketing people took the old products and tarted them up. It was a terrible policy problem, our image was suffering badly; we were frittering away our resources tinkering with old products and it led to confusion in the marketplace over who we were.'*
>
> **Vic Lilly**
> **Kango**

The best approach may be to involve design consultancies in commissioning and analysing the research relevant to your design project. Obviously there will be some findings which suggest basic do's and don'ts – for example, the colour red on toothpaste packaging would be very unacceptable in France – but

Styling

Foam models were produced by industrial design consultancy Grey Matter to show three possible versions of the same product, an electric can crusher. From left to right: the 'functional' version, a combination of moulded and sheet metal components, would be mounted on a door or cupboard; the mainstream 'consumer product', housed in moulded casework, reflects current trends in kitchen appliances; and of the 'high style' version, designer Kevin Thompson of Grey Matter says it was designed to show that 'form does not have to follow function - form can follow emotion'.

Once the general stylistic direction of a product has been decided, it can be reinforced with colour, detailing and surface finishes, product graphics, packaging, accompanying literature and point-of-sale material.

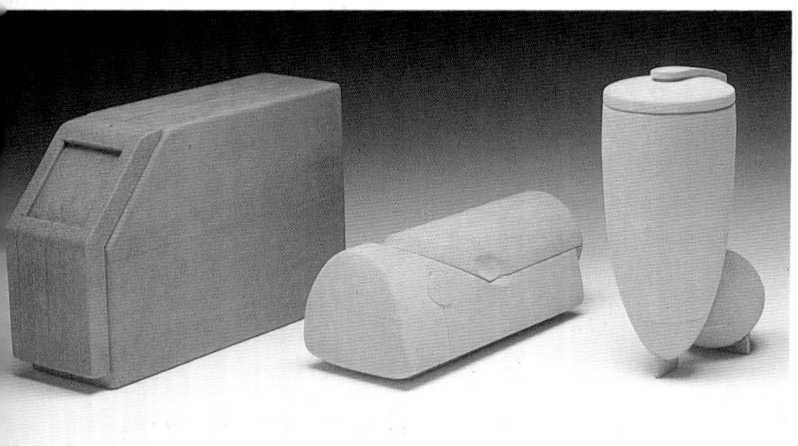

Mood

'Mood boards' are variously used by designers to plot trends in colour, shape and texture, and to deduce possible stylistic directions for design projects. These examples were produced by industrial design consultancy Queensberry Hunt. The 'country cottage' images are supplemented with a list of relevant materials, colours, and retailers specializing in that 'look'. The exotic/tropical version is left to speak for itself.

Design consultancies will either use their own observations to compile these boards or will commission appropriate research from forecasting bodies, which are most numerous in the fashion and textile industries. Their influence extends into many other areas, however - the automotive industry, for example, watches fashion trends closely.

there should be some flexibility over the interpretation of less conclusive data.

Once armed with a clear idea of your target audience and where your product or service will sit in the marketplace, you will need to explore with your designer how to achieve the market position you require. Positioning any design correctly in relation to its competitors is a key point. Clients should go to briefing sessions armed with as much information on competitive products as they can gather, and samples if possible.

Correct positioning is about presenting not only the performance characteristics which are appropriate for your target market, but about less tangible factors too – such as emotive and tactile qualities, which may be expressed in the details of your design rather than the grand plan. It is not, as one manufacturer thought, a matter of painting something red, white and blue to establish it as British and high quality.

Styling

Styling offers clients many options. Products, environments and communications can be styled in multifarious ways: for example high quality, cheap and cheerful, mass market, elite, 'masculine', futuristic and so on. Designs can be created feature-rich and stripped down to various levels of simplicity to be marketed at different price points.

There has been a lingering distrust of styling in some quarters, but Sony's successful marketing of more than 300 different versions of the basic Walkman has no doubt been influential in changing people's minds. Indeed, such is the new respectability of styling that some consultancies with leading edge skills in electronics and engineering play down these skills in order to play up their styling expertise.

Aesthetic appeal is clearly not the only characteristic of good design and is much less important in some areas than in others. Yet there are areas where design has historically not been considered important – medical products, agricultural equipment – which have gained immeasurably from some styling input. JCB with its back-hoe loaders, and Maxview with its revolutionary aerials, have proved that styling can be as lucrative for unglamorous products as for fashion-sensitive consumer goods.

Materials and processes are influential in interpreting any design, and some designers will undertake to source materials and manufacturers or printers. Graphic design projects should be the easiest to control, since they usually involve known materials and technology, but mistakes still abound: urgently needed fliers printed using slow-drying ink; direct reply cards printed on paper lighter than the weight required by the Post Office.

Some designers will be very knowledgeable about materials, but it has to be borne in mind that most will not. Research by the DTI in the mid-1980's found that half of 600 chief designers they interviewed knew little or nothing about the applications of nylon, and that the situation was getting worse: young designers were even less likely to specify new materials than their older counterparts. As for companies, in the UK the average company introduces a change of material every seven years (compared to every three in Japan). Materials technology is a fast growing field which presents a clear opportunity for smaller companies to compete with larger ones. Information and advice is available from several sources (see Addresses, page 85).

Colour

Colour, too, is an oft-neglected and misunderstood area, specified at the last moment usually on fairly subjective grounds. In fact, choice of colour can be a vitally important decision and unexpectedly tricky to implement – the colour you choose from a printed paper sample may not translate well or accurately to a different medium, such as plastic or fabric. The Chartered Society of Designers runs a forum called the Colour Group which is available for consultation and which can recommend sources of expertise on colour (see Addresses, page 85).

Among the many exigencies of the Single European Market is the dramatically increasing need for businesses to meet or exceed relevant international standards, including guidelines and directives on the environment. Many companies have been slow to address environmental issues and will face increasing pressure from political sources and from the public to do so. You will also probably find that it is a concern of your designer.

Respect for the environment is a principle enshrined in the Chartered Society of Designers' Code of Professional Conduct (see page 75). In fact, environmental awareness has presented many companies with the opportunity to launch new products or to relaunch or diversify existing ones. Far from being a hairshirt, environmental awareness can mean increased profitability.

Clients commissioning office, factory or shop designs will need to take account of far-reaching new legislation on workplace environments. Employees in European countries already enjoy stronger rights over the conditions in which they work. In May 1990 the European Commission adopted a directive on safety and health for people working continuously with screens, for example; it will come into force in December 1992 and will apply to all new equipment installed after that date. We can expect to see legislation relating to vapours from furniture and carpets, ozone from photocopiers, viruses from airconditioning, back strain from chairs, eye problems and repetitive strain injuries from VDUs – all of which have major implications for the infrastructure of commercial buildings. The interior design of offices is an increasingly specialized business. As Philip Ross of Business Design Group observes: 'Anyone who thinks that office planning is about arranging the furniture is in for a rude awakening'.

Multi-discipline projects

Many design projects are cross-disciplinary. A product design project may involve product graphics, packaging and instructions. A graphic design project may involve elements of three-dimensional design, such as signage or structural packaging. Interior design projects often contain elements of product and graphic design: perhaps special lighting, and menus. Lines of communication and responsibility between all designers and contractors involved should be made clear in your brief. An interior design project, for example, may involve an architect and a quantity surveyor as well as an interior designer and graphic designer.

In fact, there is a general perception that each design discipline is more self-contained that it really is. People tend to think of product design as a two-dimensional drafting exercise, when in fact it encompasses (or should encompass) decisions on styling, materials, ergonomics, marketing, safety, and packaging – both structural for safe transportation and with adequate user or point-of-sale instructions. (An astonishing number of products come off the production line without anything to put them in.)

Another factor to bear in mind may be the prevention of vandalism, or at least the minimizing of wear and tear. It is a little-known fact that the housings of public telephone boxes are regularly subjected to fire, gouging with bread knives, butane gas played over the coin runway, attack by solvents and acid, and balloons inserted and inflated inside the coin safe. Telecommunications company Mercury only released its latest housings after a specially commissioned team of safebreakers had failed to destroy it.

BRIEFING METHODS

Conventional wisdom has it that a written brief is the best insurance, but when clients don't know what is possible or what the options are, the design brief is arrived at through a process of negotiation with the designer.

Unusually among design management theorists, Angela Dumas, director of the Centre for Design Management at the London Business School, is an advocate of verbal briefings. She suggests that briefs be worked out through a series of discussions which are recorded in order to form the basis of a contract.

'Both managers and designers fall back on the existence of a written brief as a form of reassurance that everything is clearly understood on both sides. In an ideal world, this might be the case, but experience shows that often both sides take away a different

understanding of the written brief. The formal written approach to briefing narrows down the opportunities for an open-ended exploration of the problem and consequently reduces the opportunity for an innovative solution.'

Angela Dumas , London Business School

Designers often present their first rough concepts in the form of 'theme boards' (or 'lifestyle boards' or 'mood boards'), collages of images to show the kind of shapes, colours and finishes which could be relevant to the project. Possibly a name or phrase will sum up the mood. There is no reason why clients themselves shouldn't adopt this method to communicate ideas to designers.

When running design management courses, Dumas uses slides, magazines, words and objects which she encourages managers to analyse and, where appropriate, handle, take apart and redesign. This helps people to appreciate the qualities which give each design its uniqueness. It also helps them to develop a vocabulary with which to describe design attributes, and thus to define new design concepts.

For products aimed at architects, interior designers, and specifiers, industrial designer Ben Fether at FM Design has produced a briefing aid in the form of a cake diagram, in which all current aesthetics in buildings and related products are identified in separate slices of the cake. Thus he illustrates a range of possible directions for a design to take and makes it easier for the designer and client to share a common visual language when agreeing the approach.

Another way of establishing a visual reference point is to devise an appropriate metaphor. The Honda product team in the US did this when developing the Accord. The concept, fairly abstract, was: 'a rugby player in a dinner suit'. Once fixed, this image was able to integrate the various design inputs to the car. Whoever designed the door handle or the dashboard designed it with that image in mind. In the UK, Clarks shoes have done something similar. The inspiration for one shoe was the front end of a Japanese train. Says Angela Dumas, 'The level of detail achieved in design projects through this sort of exercise is always significantly higher than those developed in the usual way.'

AROMA: A BRIEFING CASE STUDY

❝ Michael Zur-Spiro is a former management consultant who, together with his wife Susan, decided to launch a chain of small food outlets in London. They had never used a design consultancy before, although their experience of marketing meant that their approach to design was relatively sophisticated. Nevertheless, the project encapsulated some classic problems.

'On the continent there are a number of eating places based around coffee and food. They're not like English cafés, which are either places you can have lunch or just a sandwich; they are something in between – easy places to go into for a casual, fifteen-minute consumption of something excellent. I wanted something which would fill that gap in London, something which would also cross food and retail to induce impulse buying of additional products.

I put together a business plan. I did research into the continental café chains and gathered a lot of information on the economics of those firms: of two large German chains in particular. I needed to raise money from outside sources and wanted to do a very professional job with the business plan. I thought it would be good to have some drawings to go with it. I was friendly with partners in an architecture firm, ORMS, and they introduced me to ADC [Associated Design Consultants], a graphics consultancy.

Neither ORMS nor ADC had done the kind of planning work I had in mind, but together we worked out how to illustrate the business plan. They both did the work for free – it was a hell of a lot to do for nothing. ORMS did the layout and ADC the graphic images.

Implementation

*Michael Zur-Spiro in his successful café/shop, Aroma.
Commissioning a design consultancy was a new experience for
him, and the larger issues (such as establishing the corporate
identity, planning the flow of people through the shop and
organizing the furnishings and wall finishes) were unexpectedly
easier to deal with than the small details.*

Packaging

Packaging the coffee for Aroma seemed a simple project but proved unexpectedly full of design problems. Gas contained in the coffee bags meant that the boxes ordered were fractionally too small. There were choices to be made about the type of finish specified for the paper. The rubber stamp used to identify types of coffee, though an effective and cheap solution to labelling, was difficult to use at speed without smudging the ink.

The presentation I then gave to the venture capital company was almost too slick; the design work actually caused problems for me. It turned out that neither ORMS nor ADC had really understood the Aroma concept properly, and thought of it as a sleek bar with lots of chrome. In the briefing document, however, I had talked a lot about Mexico and sunshine. When people come into the shop I want it to be like a "fifteen-minute holiday in the sun"; that was the briefing slogan if you like. I couldn't ask for any more work from the consultancies as they had done so much for free, but the venture capital companies found some inconsistencies between the wording of my proposal and the drawings. However, I raised the money and I got help from the DTI's Enterprise Initiative.

Before launching the actual project, Susan and I decided not to continue with ORMS and ADC automatically but to choose the right professionals for the work. We carefully interviewed a small number of architects and design consultancies and went through a fairly rigorous selection process. We had prepared our own boards to show the images, feelings and colours we had in mind.

The choice of architect was more straightforward than the choice of graphics consultancy. The differences between graphics groups were subtle and we had no way of predicting what they would each come up with, since our corporate identity was a pretty abstract concept at that stage. In the end we went with ADC, as its creative director Nick Hockley's proposal was concise and imaginative, and we had an excellent rapport with him. The disadvantage of ADC was that they had not done food packaging before, and this did cause a few problems later on.

We thought that the design fees would be markedly lower than they were. Even after ADC reduced their fees to match lower quotes, it still cost us a great deal of money!

There was an excellent month which was concerned with the

corporate identity. For the logo I had been thinking of something whimsical, a bit like the way "España" is written for the Spanish tourist board, a bit Miróesque. But ADC took the "O" from Aroma and chiselled it out into a 3D form and then shone a light on it, and it produced this modern, stark logo which looked almost like cyrillic script – very bold and graphic – which by now I absolutely love. But at the time I said "What are the alternatives?" I thought it was too cold but they just asked me to live with it for a while, and after a week I was smitten with it. I no longer think it's cold. It is a stylized sun, and you can also see lots of other things in it – the swirl of coffee; night and day.

That was the fun bit. Stage two was not fun. ADC had undertaken to identify suppliers of paper products: paper plates, paper cups, boxes for the coffee, serviettes, sandwich containers and so on. All of the packaging aspects were a key part of the project, but it seemed to me that ADC's interest had flagged. Things got really tense. I had the shops ready and we couldn't get a firm recommendation out of ADC about these paper products. I now think that they had never dealt with suppliers of these products before, although at the time I thought they had. We had a lot of acrimony because I kept expecting things on certain dates and didn't get them. Correspondence between us proves that we set the situation up badly.

One thing we really wanted and got exactly right was the articulation of colour and sunniness. We wanted all of the packaging to fit into our colour scheme, but our choice of packaging – for the coffee, for example – was limited by our budget. We thought of using coffee bags but ADC thought that would look too messy. Personally I think they might have looked messy but they would also have looked more appealing, more tactile.

ADC wanted us to put the coffee in boxes and came up with the economical idea that different coffees could be distinguished by different rubber stamps. They suggested grey boxes, but that did not fit into our scheme at all – grey had no place in Aroma! We thought of using a strong yellow and the consultancy thought our idea was great!

We liked ADC's ideas but we had all sorts of problems, some of which were resolved badly and others which we have simply had to live with. We found a printer to produce the boxes; he wanted to know whether to make them in coated or uncoated paper. Naturally we asked ADC which to use but it took a good three weeks before they let us know. They advised coated paper; they had apparently tested and found that the rubber stamps could be used on coated paper but in practice it has proved tremendously difficult to stamp the boxes without smudging the ink. This has given the shop staff a big problem.

We also hadn't reckoned on the coffee bags containing some gas, so the boxes we ordered turned out to be a bit small. Until this batch of boxes is used up, we have to squeeze the gas out of each bag and stuff it into the box!

The final phase was post-opening. The shop works well, but the layout is sub-optimal. The retail products – brightly coloured crockery – don't sell well because they are mostly tucked away behind counters.

ADC had also promised me a resource book, something that would tell me everything I needed to know about the logo, the typeface, the carrier bags and so on. When it finally materialized it was two months late and useless. It contained factual errors. It had been given to a junior person – fair enough – but that person simply hadn't understood the project. The result was that I had to take on more myself. I understand that they don't want to be chasing after paper cups, but on the other hand I had bought that time.

We have had some major problems and, if I may say so, there is a real lesson to be learnt from my experience. We should have stated our expectations more clearly, being very specific about the output we required.

With the internal signage, for example, we had always assumed (and thought we had discussed) that ADC would provide a shelf signage system, and various hanging signs. We kept asking for them but ADC said they were finished with the work! In the end we made the signs ourselves on an Apple Mac and to be frank, they don't look very good. We feel they should have told us at the outset we would need internal signage in the shop.

On the other hand, Nick's view is that "Graphics is about finding ways of not putting up graphics". For example, I had said that we needed a sign saying "Please pay here". Nick said not to do this but instead to move the cash register. We did and it was astounding how much better the shop flowed. He was tremendously thoughtful about the customer and he has a very good understanding of the interplay between graphics and space, which is a difficult thing to get right.

We wrote a good brief but we missed some things out. One thing I would suggest to someone using design for the first time is to think carefully about what is wanted and to write it down, but it could make things rigid. We're doing things differently for the second shop. For a start we're more experienced so we know better what we need and how to get it. We've also contracted Nick Hockley to work with the company for a fixed number of hours per week for a couple of months, and he'll be working with me almost as part of Aroma, as a business partner.

Should you use a designer at all? We did sometimes wonder whether we could have managed ourselves. But frankly there is no way you should even consider doing something like this without one. The people from the venture capital firm said to me that whatever I did, I should make sure that the concept was carried through from the products to the menu to the shop. You need someone to guide you through that. You need someone to think deeply about the brand.'

Maxine Horn:, a director of Associated Design Consultants, puts her side of the story.

'I know exactly the sort of gripes he has. But from our point of view he sometimes expected too much, wanting more but not prepared to pay for it. At times he asked us, almost as a favour, just "to think about some signage" long after the time he had paid us for was exhausted. He paid us £15,000, which is half what we originally quoted (to take into consideration his total requirement); £4,000 of that he had received from the DTI. Overall fees amounted to £40,000. So we made a loss of £25,000, and he effectively got the job for £11,000. He felt that it would be an investment for us because as the company grows and there is a roll-out programme we'll get the work and will reap the benefits then.

But occasionally he went over the goodwill threshold, treating too many requests as favours and never putting instructions in writing or on any basis on which we could expect to get paid.

The most frustrating aspects were when he complained about speed of delivery (of the "favours") or poor quality of, for example, carrier bags from a manufacturer we had suggested. In fact, we had originally suggested a higher quality carrier bag but he said it was too expensive.

Fortunately, despite the money wrangles, we got along well. I just hope he remembers our "investment" and that we will indeed "reap the benefits" in the future.

5
legal issues

5 SAFEGUARDING YOUR INTERESTS

Clients should be aware that there is potential for litigation from three sources: from their design consultancy (for example, for breach of contract), from competitors (for example, for infringement of copyright) and from end users (for example, for damage caused by a product).

Legal action by clients against their designers is also becoming common. This partly reflects changes in legislation. Before the 1987 Consumer Protection Act, for example, consumers suing for negligence had to demonstrate that manufacturers had been negligent; now they only have to show that they have an injury. Inadequate packaging or ambiguous instructions can also constitute grounds for negligence. Design in all its forms is increasingly vulnerable to litigation. Many designers belong to a professional indemnity scheme which gives them some protection.

There is no commitment, financial or otherwise, between client and designer during briefing and fee negotiations. Once a proposal has been agreed to and signed by both parties, however, this constitutes a contract to which either party can refer in case of need. It is a good idea to have your contact looked over by a solicitor conversant with contract law.

Keep written notes of meetings; most designers do the same. It may be helpful to swap these 'contact reports' as a matter of routine. Get written estimates from the consultancy for any ad hoc changes to the project as it develops.

Your finished design may be acceptable, not acceptable, or acceptable with amendments. If it can be shown that the designer has not met the brief, then the designer should undertake to resolve the problems at no further cost to the client.

If the work does meet the brief but the client does not like it, any further work will be for a negotiable fee. In cases of dispute, the Chartered Society of Designers can recommend an expert who will give a view on the standard of the work. Before reaching this point, however, it may be helpful to look at the development work done by the designer. Seeing the process which has led to the

Control

**Once you have corporate design specifications in place, one way
of making sure that everyone in your business understands them
is to produce a design manual. In recent years there has been a
trend away from huge, all-encompassing manuals, which tended
to sit on shelves gathering dust, in favour of slimmer volumes
explaining the general principles to be observed. This manual,
produced by London Underground, was well received.**

proposed solution can sometimes reassure clients that the work, while not perhaps what was expected, is the logical result of a thorough analysis of the brief.

Designers are likely to sue clients who do not pay their bills, though it may be possible to negotiate payment by instalments. There have been cases in which clients who could not or would not pay designers' bills have claimed that the design work was not up to standard, and in these instances an independent verdict would be sought.

After any design project is completed, it is common for designers to ask for examples or photographs of the work to keep in their own portfolios. Designers must get permission from their client to hold such material, and are bound not to reveal sensitive information about the design.

A client can ask a designer not to work for another company in the same market, in a similar business or geographical area or for a period of time, but this arrangement will usually cost money. Most designers will agree to what they consider reasonable requests, though these may be difficult to enforce by law.

RIGHTS

Any form of rights infringement, even if committed unwittingly, can invite legal action forcing the withdrawal of the 'guilty' design. Beware of asking designers to produce a design similar to one which already exists: you and the designer could be sued for infringement of copyright. There have been some notable cases relating to own-brand packaging aping the packaging of market leading brands. If you are in any doubt about the likelihood of copyright infringement, consult a patent agent or a lawyer who specializes in copyright law.

Most design consultancies are not sufficiently expert in design protection law to advise clients on which rights are most appropriate for them, nor on whether a proposed design will infringe existing design rights. Designers may advise clients that they cannot help on such matters and may escape liability in the event of litigation, but the clients will not.

There are two automatic rights which come into being without any form of registration.

Copyright

Under the Copyright, Designs and Patents Act 1988, copyright automatically belongs to the person who creates the work (ie the designer), and it protects original, literary, dramatic, musical and artistic works, sound recordings, films and broadcasts, and computer programs. In the context of design, copyright includes technical descriptions and engineering drawings, graphic design, photography, illustration and copywriting. To be protected by copyright the work does not actually have to be new or even have any aesthetic value; the work just has to be the result of independent intellectual effort.

Copyright cannot be held in a name, a title, or an idea, but in its original realization or manifestation.

Clients should acquire the copyright in all design work once they have paid for it; if the designer wants to retain or limit the use of the design, a lower fee may be negotiated.

The designer usually keeps copyright until the final stage of the project has been completed and fees have been paid. Sometimes the designer will retain copyright on unused ideas. Where a design consultancy uses bought-in services, the consultancy should ensure that any subcontractors assign copyright of their work to the client on payment.

The copyright owner has the right to control the ways in which the work is exploited, including copying, adapting, issuing copies to the public, performing in public and broadcasting. The 1988 Act also created 'moral rights' which allow the creator of a work (who may not necessarily be the copyright owner) the right to be identified on the work.

Copyright usually lasts up to 50 years after the death of the author/creator, but there are exceptions. In all cases it is advisable to use the services of a copyright lawyer to ensure that potentially litigious situations are avoided.

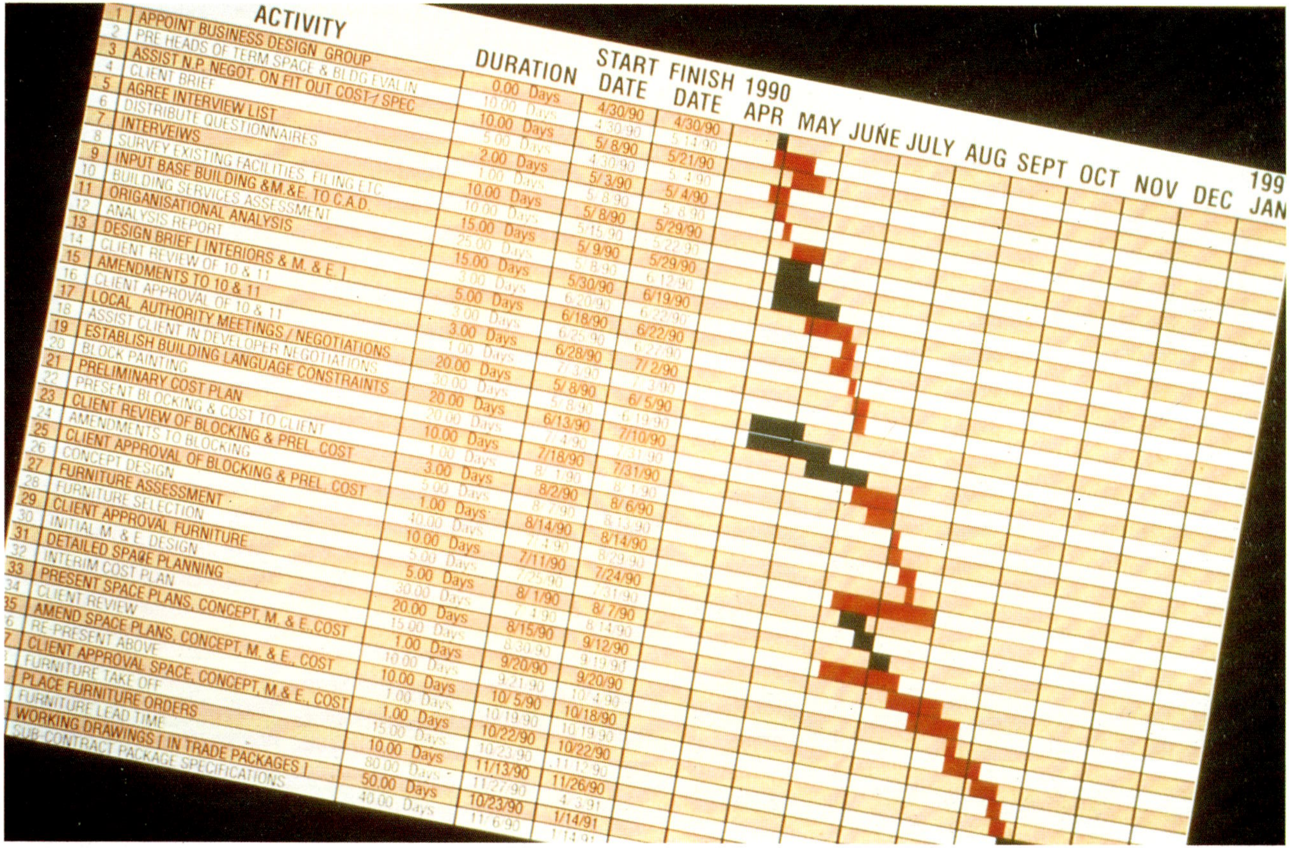

How to Buy Design

The photographed critical-path (Gantt) chart carries the following table. Its timeline scale runs across the months **APR MAY JUNE JULY AUG SEPT OCT NOV DEC JAN** (1990–1991).

No.	ACTIVITY	DURATION	START DATE	FINISH DATE
1	APPOINT BUSINESS DESIGN GROUP	0.00 Days	4/30/90	4/30/90
2	PRE HEADS OF TERM SPACE & BLDG EVAL IN	10.00 Days	4/30/90	5/14/90
3	ASSIST N.P. NEGOT. ON FIT OUT COST / SPEC	10.00 Days	5/8/90	5/21/90
4	CLIENT BRIEF	5.00 Days	4/30/90	5/4/90
5	AGREE INTERVIEW LIST	2.00 Days	5/3/90	5/4/90
6	DISTRIBUTE QUESTIONNAIRES	1.00 Days	5/8/90	5/8/90
7	INTERVEIWS	10.00 Days	5/8/90	5/29/90
8	SURVEY EXISTING FACILITIES, FILING ETC	10.00 Days	5/15/90	5/22/90
9	INPUT BASE BUILDING &M.&E. TO C.A.D.	15.00 Days	5/9/90	5/29/90
10	BUILDING SERVICES ASSESSMENT	25.00 Days	5/8/90	6/12/90
11	ORIGANISATIONAL ANALYSIS	15.00 Days	5/30/90	6/19/90
12	ANALYSIS REPORT	3.00 Days	6/20/90	6/22/90
13	DESIGN BRIEF (INTERIORS & M. & E.)	5.00 Days	6/18/90	6/22/90
14	CLIENT REVIEW OF 10 & 11	3.00 Days	6/25/90	6/27/90
15	AMENDMENTS TO 10 & 11	3.00 Days	6/28/90	7/2/90
16	CLIENT APPROVAL OF 10 & 11	1.00 Days	7/3/90	7/3/90
17	LOCAL AUTHORITY MEETINGS / NEGOTIATIONS	20.00 Days	5/8/90	6/5/90
18	ASSIST CLIENT IN DEVELOPER NEGOTIATIONS	30.00 Days	5/8/90	6/19/90
19	ESTABLISH BUILDING LANGUAGE CONSTRAINTS	20.00 Days	6/13/90	7/10/90
20	BLOCK PAINTING	20.00 Days	7/4/90	7/31/90
21	PRELIMINARY COST PLAN	10.00 Days	7/18/90	7/31/90
22	PRESENT BLOCKING & COST TO CLIENT	1.00 Days	8/1/90	8/1/90
23	CLIENT REVIEW OF BLOCKING & PREL. COST	3.00 Days	8/2/90	8/6/90
24	AMENDMENTS TO BLOCKING	5.00 Days	8/7/90	8/13/90
25	CLIENT APPROVAL OF BLOCKING & PREL. COST	1.00 Days	8/14/90	8/14/90
26	CONCEPT DESIGN	40.00 Days	7/4/90	8/29/90
27	FURNITURE ASSESSMENT	10.00 Days	7/11/90	7/24/90
28	FURNITURE SELECTION	5.00 Days	7/25/90	7/31/90
29	CLIENT APPROVAL FURNITURE	5.00 Days	8/1/90	8/7/90
30	INITIAL M & E DESIGN	30.00 Days	7/4/90	8/14/90
31	DETAILED SPACE PLANNING	20.00 Days	8/15/90	9/12/90
32	INTERIM COST PLAN	15.00 Days	8/30/90	9/19/90
33	PRESENT SPACE PLANS, CONCEPT, M. & E. COST	1.00 Days	9/20/90	9/20/90
34	CLIENT REVIEW	10.00 Days	9/21/90	10/4/90
35	AMEND SPACE PLANS, CONCEPT, M. & E. COST	10.00 Days	10/5/90	10/18/90
36	RE-PRESENT ABOVE	1.00 Days	10/19/90	10/19/90
37	CLIENT APPROVAL SPACE, CONCEPT, M. & E. COST	1.00 Days	10/22/90	10/22/90
38	FURNITURE TAKE OFF	15.00 Days	10/23/90	11/12/90
39	PLACE FURNITURE ORDERS	10.00 Days	11/13/90	11/26/90
40	FURNITURE LEAD TIME	80.00 Days	11/27/90	4/3/91
41	WORKING DRAWINGS (IN TRADE PACKAGES)	50.00 Days	10/23/90	1/14/91
42	SUB-CONTRACT PACKAGE SPECIFICATIONS	40.00 Days	11/6/90	[illegible]

Planning – A critical path chart, compiled by interior design consultancy Business Design Group, shows that there is more to office design than meets the eye. Especially critical are the 'client approval' phases, which should always be built realistically into any design schedule. Changes of mind on the part of the client are probably the major cause of delay in all design work.

Unregistered design right

Unregistered design right automatically belongs to the person who commissioned the design (ie the client), and it relates to the original shape or configuration of the work. It only applies to work in three dimensions. There is no need to register the design, but in case of infringement by others it is worth keeping a note of the date on which the design first reaches the market.

Unregistered design right gives the client the right to prevent any copying of the design for a period of up to 15 years. Unregistered design right can be bought, sold or licensed.

The following rights, in order to exist, must be registered by a patent agent. Such agents specialize in establishing the appropriate protection (internationally if necessary) for any innovation, development or new idea. Patent agents (see Patent Office in Addresses) can also deal with licensing rights and conflicts.

Patents

Patents can be applied for in virtually all machines, products and processes (and parts of them) in all sectors of industry, provided that they satisfy three criteria: they have to be new, inventive, and capable of industrial application. A patent gives exclusive use of an invention for up to 20 years. A patent may be bought, sold, hired or licensed.

The EC Commission estimates that European industries waste more than £20 billion annually by repeating the efforts of others – easily identifiable through a search of published patents. The Patent Office offers a Search and Advisory Service which, for a fee (£200–£400 at 1992 prices), can retrieve information on UK, European, US and world patents. As well as revealing whether any given project has a precedent, such searches can be valuable in terms of general information-gathering. They can indicate, for example, which new markets competitors are developing and in which countries.

The Search and Advisory Service also offers (among other things) patentability assessments, name searches, and a 'state of the art review', a compilation of patent documents providing an overview of a specific field, or on the most recent developments in a given area of technology.

Registered design right

Registered design right relates to the outward appearance of a manufactured article or set of articles. To be eligible for registration, designs must be new, and their aesthetic appeal must be considered relevant. Purely functional designs are not registerable; nor are the parts of a design which are dictated by the shape of the whole. Modifications to previous designs can be registered.

Registered design right lasts for five years initially and can be extended up to a maximum of 25 years. A registered design can be bought, sold, hired or licensed.

Trade and service marks

Trade and service marks are identification symbols used to distinguish one trader's goods or services from another's.

EUROPEAN RIGHTS

European law on design protection is still evolving. Variations on all of the above forms of protection and differences in interpretation exist throughout Europe. National design protection systems will continue for the foreseeable future, but the hope is that the cost and time benefits of EC protection will ultimately result in the redundancy of national systems.

All EC countries, except Greece, have some form of registered design right; a Community Patent and a Community Trade Mark are in progress. These are developments watched with interest by designers. The EC defines a trade mark, for example, as any sign capable of being represented graphically: a definition which will allow the shape of packaging to be a trade mark for the first time in the UK.

Although some aspects of copyright will be adopted throughout Europe, it has been thought too complex to be fully harmonized by the 1 January 1993 single market deadline. It is therefore proposed that an alternative community design system will automatically become law throughout the EC. There will be two forms of protection. Disputes over the validity of registrations will be referred to the European Court of Justice.

Registered Community design right

Registered Community design right belongs automatically to the designer, or to the client if this has been negotiated with the designer. It protects designs 'of distinctive character' that must be unknown to specialists in the sector concerned, and recognizably different from existing designs. This relates to appearance only, except where design is dictated by function. This right prevents unauthorized copying for up to a maximum of 25 years. A fee must be paid to a Community Design Office.

Unregistered Community design right

Unregistered Community design right belongs automatically to the designer, or to the client if this has been negotiated with the designer. It protects the same categories of design as registered community design (see above), but the unregistered version is generally intended for industries producing designs in high volume, not all of which will reach the market. The right prevents copying for a period of up to three years. No fee is involved and no form of registration is necessary.

6 the design process

6   TIMESCALES

Whatever the project, the design process is likely to follow these basic steps:

- **agreeing the brief**
- **conducting any necessary research**
- **producing first rough ideas**
- **shortlisting ideas**
- **specifying materials**
- **presenting ideas in a more finished form**
- **testing if necessary**
- **production**
- **evaluation.**

It is very difficult to generalize about the length of time this process will take on any given project; there are so many variables involved. Timescales will depend on such factors as the research, planning and testing needed, the strategic importance of the project, its technology content, the experience and workload of the consultancy, the speed with which you are able to approve each stage of the design work, and how much you are prepared to pay for fast turnaround – design consultancies may be prepared to work evenings and weekends in return for rush rates.

Nevertheless, it may be useful to give some very broad brush timescales. For example, a single item, such as a small task lamp, a single garden chair, a security device, a basic kiosk/snack bar or a small piece of packaging might be ready for production in under four weeks. A larger project, such as a range of task lamps, an interior domestic chair, the layout of a car dashboard, a range of packaging, an annual report, a small shop or restaurant, might take between one and six months. A major project such as a complete lighting or office furniture system, a car, a large shop (for example, a DIY shed), a museum, hotel, or corporate identity programme would be likely to take a year or more.

Design projects cry out for planning. The design process is full of interdependent decisions: apparently minor changes can have major cost and timescale implications, and the later those changes are made, the more expensive the implications will be. Schedules for design projects often fail to allow enough time for each stage to be cleared by all of the people who have to approve it before it can proceed. Where efforts are made to telescope the design process, as in simultaneous engineering, juggling the variables of cost and time is even more demanding. All complex projects will call for some form of critical path analysis, and even simple projects need planning if they are to result in the best quality at the cheapest price in the shortest time.

The chart on page 64 shows the planned progress of an office interior design project, with critical approval phases indicated. Looking at this makes it easier to understand why a two-day delay in approving a particular phase of the project can add six weeks to the final delivery date.

Not all projects are as complex as this (some are more so) but all design projects do consist of a web of interrelated actions. Active communication between all parties involved is very important in keeping to schedules.

DESIGN MANAGEMENT

Ideally, all businesses should have at least one person whose responsibility it is to maximize the benefits of design in the organization. Alas, few businesses do. Most design projects are commissioned on an ad hoc, jobbing basis, and the results of the work are rarely quantified.

In this respect little has changed since the 1988 survey of the design consultancy sector by investment analysts James Capel, which concluded that the management of design was in general 'erratic and unplanned'.

There are many people whose job includes commissioning design – for example the thousands of people who commission print work every year – but usually they are not encouraged to see this as part of a larger picture. In due course they move on to other jobs, accumulated design knowledge is lost, and relationships with particular designers or consultancies fall into disrepair. In some companies several people may be commissioning design services, unaware of the others' actions. It adds up to a picture of wasted effort and missed opportunity.

The companies which deal most successfully with design tend to have clearly identified design budgets, clearly designated design managers, clear cross-disciplinary communication, and evaluation procedures which feed back into the design process.

Where resources allow it, having a designated design manager (or managers) is half the battle. At a practical level, it provides a single point of contact for external design suppliers. More fundamentally, he or she has to 'own the problem' – without which little progress can be made. BT for example, has dedicated design management teams which co-ordinate the efforts of external designers, in-house designers, and task forces of senior staff assigned to projects as necessary. Without such an infrastructure, a design programme as extensive as BT's would not only be impossible to control, it would not have been initiated.

Design managers don't necessarily have to be *au fait* with design trends. What they do need is support from the top of their organization. Design often forces a radical reappraisal of the way things are done and without senior support many design projects fall at the first cavil. The design management seminars run by the Chartered Society of Designers in 1991 concluded that design managers should aim to establish a 'powerbase', from which design can be sold effectively to the rest of the organization.

Most design consultants will take a responsible, not to say protective, attitude to their clients, and will see it as part of their job to make sure that everything goes as it should and that you consistently have a good story to tell your colleagues.

Almost inevitably, though, the process of creating something new while working with designers you may never have met before has potential for trouble.

You may encounter attitude problems – perhaps the designers seem more interested in acclaim from their peers than in your needs – or simple lack of ability. But this is rare. As Tony Key, corporate head of design at BT, observes: 'I believe that the vast majority of designers are honest and sincere, and that they sell me design of very good quality at reasonable prices.'

More common are complaints from clients that they need from their designers a better understanding of business issues, better cost control, and a more detailed billing system. Lack of understanding of production processes, either manufacturing or printing, can be a problem, as can flagging interest on the part of the designer when the first flush of creative problem-solving is succeeded by routine servicing of the project.

> *'With graphic designers we find that they sometimes lack understanding of and sympathy with the less glamorous design requirements. In the past we have been very pleased with initial packaging design concepts for ranges of products but found that the individual products in the range have looked too similar. I have also found that designers can create designs that look good in the hand but much less effective on the shelf. We have been partly to blame for not being sufficiently clear in our briefing, but when you're paying for an experienced consultancy you expect them to be aware of these aspects.'*
>
> **Steve Hallam**
> **Fisons**

Generally, these problems can be avoided through close contact and communication with your designers. Some clients want to be involved in every stage of their design project and want to see every option along the way; others feel that what ends up on the studio floor is none of their business. On balance, keeping in close touch is advisable. It helps the designer to stay in tune with what you want and it helps you to report on the design process to colleagues with greater confidence.

> *'Recently we handled two projects, one by me and one by a colleague. They proceeded simultaneously and there was a lot of work which was common to both, and both used the same design consultancy. There was a lesson in the process. The problems lay in communication: I'm used to dealing with designers and I refer back, making sure it's all going as I want it. My colleague's understanding was*

that he should write a brief, give the job away and wait for the result ... the design consultancy itself was a bit remiss in not communicating more with us, but it was six of one and half a dozen of the other.'

Bill McMichael
Xpelair

Angela Dumas believes that design management is influenced by the type of organization in which it is practised. Part of her design management education programme at the London Business School involves a test exercise in which managers in three groups, from entrepreneur, bureaucratic and political organizations, are asked to role-play the planning and commissioning of a new restaurant. The results, Dumas says, are consistent. By the end of the allotted timespan, the entrepreneurs have always invited the designer in to work alongside the management team, and have produced highly detailed design proposals including what sort of food will be served and how it will be presented. The bureaucracy will have developed some bland general proposals and a lot of rules and regulations about design, but nothing specific. The political organizations will have got almost nowhere in the same amount of time. The moral – that bringing in the designer as a business partner is best – is clear.

Designers have their complaints about clients: most commonly that clients change their minds too often for no good reason, are literally inaccessible at vital stages in the process, are too conservative in their choice of design, are slow to approve work or to pay for it, or abuse copyright agreements.

But most client/designer relationships survive whatever traumas ensue from their joint enterprises. Keith Lawson of Business Design Group describes the liaison: 'Like a brief marriage – it must be able to withstand a falling-out'. Most design consultancies, in fact, have a high proportion of repeat business.

There is clearly a balance to be struck between respecting the designer's skills and sticking to your own agenda. On the one hand you must be prepared to believe that their judgement will exceed yours – if you're paying experts, allow them to influence you. On the other, you should not be afraid to reject design work that is patently inappropriate, or to complain if you feel standards of service are slipping.

As you become more experienced as a design buyer you will become increasingly adept at this, as you will also at finding the most appropriate designers for given projects. British Rail's InterCity 250 train (due to run in 1995) had its interiors, seating, toilets, livery, cockpit, windscreen and front lights (among other things) created by design consultancies with no previous experience of trains or of mass transport design in any form. British Rail's design managers were able to judge that the consultancies commissioned – Seymour Powell and FM Design – had the requisite fundamental skills in ergonomics, aesthetics and aerodynamics to turn their hands successfully to the job. Much of the satisfaction in design management comes through creative matchmaking of designers and projects.

Ironically, this is perhaps easier for the lone entrepreneur or the person low down in the corporate hierarchy than it is for senior managers in large organizations, whose high profile makes them cautious. As Marcello Minale of design consultancy Minale Tattersfield has written: 'I have observed that the lower down the corporate ladder you go, the more adventurous the clients become.' This is, he believes, because more junior employees stand to gain if the project goes well but won't suffer unduly if it doesn't.

With so much to be gained from effective design, it would be a pity if adventurous design management were to remain the preserve of people with little to lose.

> *'The UK had the car industry in the palm of its hand in the fifties and sixties, and it threw it away by making lousy designs. The MX5 should have been an MG; it was all that the MG used to be. Nobody in the UK is making a car like that now because people thought there wasn't a market for small sports cars. British industry looks at the market where it is now, it doesn't look at where it could be if it were stimulated in an interesting way. I'd rather cater for tomorrow.'*
>
> **Roger Cooper**
> **Ideal-Standard**

APPENDIX

Extract from the Chartered Society of Designers' Code of Professional Conduct.

Professional responsibilities

Members shall conduct their business competently and act at all times with integrity and honesty.

Whilst members may publicise their services in a factual and dignified manner they shall not knowingly seek to supplant another designer already engaged on a project and if comparing their services with those of other members shall do so only in a manner which is legal, decent, honest and truthful.

Members shall treat all knowledge and information relating to their clients' or employer's business as confidential and shall not divulge such information to any third parties without the consent of the relevant client or employer.

Members shall not knowingly work simultaneously for clients or employers who are in direct competition with each other without their full knowledge.

Members shall not knowingly copy the work of another designer.

Members shall act fairly and impartially between contracting parties or when selecting others.

Members shall only enter or be a judge of competitions which comply with the design competition guidelines of the Society, or which obey rules approved by its international relations committee and when in any doubt shall consult the Director.

Members shall not subcontract the principal design work commissioned by a client or employer without their full knowledge and agreement.

Where there is any intention by a member to combine acting as a consulting designer ("consulting") with acting as a contractor including such activities as shopfitting, construction, manufacturing and printing ("contracting"), this shall be disclosed to the client and shall not be represented as one or the other alone.

Members shall have due regard to the effect of their work and endeavour that it may cause as little harm as possible either directly or indirectly to the ecology or environment, including:

living creatures

endangered species of plant or fauna

the atmosphere

rivers and seas

the land.

Members should wherever possible encourage the conservation of energy and the recycling of used products, packaging and materials.

Commissioning of members

To assist a prospective client to choose a designer, a member may present orally or in writing the following information relating to a proposed project:

a) an understanding of the brief

b) examples of the member's previous work

c) details of how the member would undertake the project

d) an estimate of fees and timescale

e) outline details of the qualifications and experience of persons to work on the project.

Before undertaking any work on a project, member shall:

a) inform the client in writing of the work to be carried out and the fees and/or charges to be paid or the basis on which they will be calculated

b) disclose to the client any interest of the member which may have a bearing on the commission, and

c) obtain the agreement of the client to pay such fees and/or charges as shall be agreed or on such basis as shall be agreed.

Members may act at their own cost on behalf of a registered charity only.

Members shall not accept payments or benefits which may impair or be construed as impairing their ability to remain impartial and fair in all their dealings.

Promotion and publicity

Members may themselves or by means of persons or organisations engaged on their behalf promote and publicise their work and their services providing always that announcements are factual, not misleading in their presentation and uphold the dignity of the profession.

Where material publicised by a member includes the work of a joint author such as another designer or an architect, the member's best endeavours shall be used to ensure that credit shall also be given to the joint author.

Members may allow their clients or employers to use the member's name in the promotion or publicising of work for which the member was responsible providing:

a) the member first approves the material which includes the member's name

b) the manner in which it is carried out is in keeping with the dignity of the profession, and

c) members may seek joint credit for work produced in their capacity as employees or as joint authors of the work when part of a team but shall not take any credit without the consent of the employer or team concerned.

PUBLICATIONS

Books/Booklets

Blackley, N (1988) *The Design Consultancy Marketplace.* London: James Capel.

British Standards Institution (1989) *Guide to Managing Product Design BS7000.* Milton Keynes: British Standards Institution.

Burall, P (1991) *Green Design.* London: The Design Council.

Department of Trade and Industry (1991) **Managing into the '90s** series:

Choosing and Appointing a Design Consultancy

Corporate Identity

Interior Design

Packaging Design

Product Design

Prepared by the Design Business Association.

Managing Product Creation

Total Product Management

Design for Effective Manufacture

Organising Product Design and Development

Managing Product Development: Six Case Studies (BS7000)

Managing the Financial Aspects of Product Design and Development

Prepared by the Design Council.

Design Council, The (1990) *Designing for Product Success: The TRIAD Design Project.* London: The Design Council.

Design Council, The/DTI (1991) *Industry and Design 1991: The Senior Managers' Viewpoint.* London: The Design Council.

Foster, R (1986) *Innovation: The Attacker's Advantage.* New York: Simon & Schuster.

Hall Richards (1991) *The Green Materials Guide for Product Designers.* London: Hall Richards.

Industrial Newsletters in association with the Department of Trade and Industry (1989) 'Simultaneous Engineering: The Management Guide 1989'.

Johnston, D (1989) *Design Protection.* London: The Design Council.

Milton, H (1990) *Packaging Design.* London: The Design Council.

Olins, W (1990) *Corporate Identity.* London: The Design Council.

Open University/UMIST (1990) *The Benefits and Costs of Investment in Design.* Milton Keynes: Open University.

PA Consulting Group (1990) *Manufacturing into the Late 1990s.* London: HMSO.

Point to Point (1991) *The Packaging Design Survey 1991.* London: Point to Point Communication.

Russell, D (1991) *Colour in Industrial Design.* London: The Design Council.

Sonsino, S (1990) *Packaging Design.* London: Thames & Hudson.

Young, P (1991) *Spy Guide to Design and Print.* London: Spy Design.

Directories

The Creative Handbook (annual)
Reed Information Services
Windsor Court
East Grinstead House
East Grinstead
West Sussex RH19 1AX
(0342) 326972

Creative Scotland (annual)
Kilmartin Publications
159 Granton Road
Edinburgh EH5 3ML
031-552 0500

The DBA Directory of Members (annual)
The Design Business Association
29 Bedford Square
London WC1B 3EG
071-631 1510

Design and Applied Arts Index (twice annually)
Design Documentation
Gurnleys
Burwash
Etchingham
East Sussex TN19 7HL
(0435) 882438

The Directory (annual)
Haymarket Publishing Services
22 Lancaster Gate
London W2 3LP
081-943 5000

The Directory of Designers (1991)
The Design Council
28 Haymarket
London SW1Y 4SU
071-839 8000

The Directory of Interior Design (every 15 months)
DID
30 Elcho Street
London SW11 4AU
071-228 3468

Magazines/Press

Blueprint (ten times annually)
Wordsearch
26 Cramer Street
London W1M 3HE
07-486 7419

Creative Review (monthly)
Centaur Publications
50 Poland Street
London W1V 4AX
071-439 4222

Design (monthly)
The Design Council
28 Haymarket
London SW1Y 4SU
071-839 8000

Designers Journal (monthly)
MBC Business Press
33-35 Bowling Green Lane
London EC1R ODA
071-837 1212

Design Review (quarterly)
Wordsearch
26 Cramer Street
London W1M 3HE
071-486 7419

Design Week (weekly)
Centaur Publications
50 Poland Street
London W1V 4AX
071-439 4222

Design World (quarterly)
Design Editorial Pty Ltd
11 School Road
Ferry Creek
Victoria 3786
Australia
(03) 755 1149

International Design (monthly)
Design Publications Inc
330 West 42nd Street
New York, NY 10036
(212) 695 4955

Databases

Creative Information Databases
13 Shorts Gardens
London WC2E 9AT
071-379 9834

Design and Applied Arts Index
Design Documentation
Gurnleys
Burwash
Etchingham
East Sussex TN19 7HL
(0435) 882438

Database on which over 200 design-related publications are indexed. Coverage of named design consultancies can be searched; enquiries welcome by post, telephone or fax.

ADDRESSES

Aram Designs Limited

3 Keane Street
London WC2
071-240 3933

Mounts annual exhibition of new design work in products
and furniture from UK colleges.

British Standards Institution

Breckland
Linford Wood
Milton Keynes MK14 6LE
(0908) 220022 Ext 2253

Of particular interest to clients may be BS5750 on quality;
BS7000 on product design management, and ISO (its
international equivalent).

Chartered Society of Designers

29 Bedford Square
London WC1B 3EG
071-631 1510

The professional body representing the interests of
designers. Among other services, the CSD publishes a Code
of Professional Conduct for designers (see page 75) and
model clauses for use as the basis of contractual
arrangements between designers and their clients. It can
also provide an independent expert witness in the event of
client/designer disputes.

Colour Group

The Studio
57 Axminster Road
London N7 6BP
071-281 7671

A forum for the exchange of colour information and trends in industrial commercial and environmental design. The group is a membership organization; membership fees range from £175 to £425.

Department of the Environment

36 Hare Street
London SE18 6L7
081-316 4416

Department of Trade and Industry

Materials Matter information line: 071-736 5339. Enterprise Initiative hotline: 071-200 1992.

Enterprise Initiative

Help under the Enterprise Initiative scheme takes the form of a free business review with an independent Enterprise Counsellor, followed by a contribution of up to two thirds of the cost of between 5 and 15 days' consultancy. Corporate identity projects are not eligible.

Design Business Association

29 Bedford Square
London WC1B 3EG
071-631 1510

The Design Council

28 Haymarket
London SW1Y 4SU
071-839 8000

The Design Council runs a Designer Selection Service, a Design Advisory Service, a Materials & Industrial Information Service, and an Innovation Business Information Service.

The Design Council also runs the Young Designers Centre, which houses changing educational exhibitions, an information service and a slide loan library.

The Design Council Central Research Unit compiles information on economic and demographic trends as they relate to design investment. In particular, it holds survey data on three areas of particular interest to the Design Council: furniture, medical products, and clothing and textiles.

The Design Council has a number of regional offices:

The Design Council Midlands

Norwich Union House
31 Waterloo Road
Wolverhampton WV1 4BP
(0902) 773631

The Design Council North

46 The Calls
Leeds LS2 7EY
(0532) 449020

Branch offices:

24th floor
Sunley Tower
Piccadilly Plaza
Manchester M1 4BA

Great North House
Sandyford Road
Newcastle-upon-Tyne
NE1 8ND

How to Buy Design

The Design Council Northern Ireland

Business Design Centre
39 Corporation Street
Belfast BT1 3BA
(0232) 238452

The Design Council Scotland

Ca' d'Oro Building
45 Gordon Street
Glasgow G1 3LZ
041-221 6121

Y Cyngor Cynllunio
The Design Council Wales

QED Centre
Main Avenue
Treforest Estate
Treforest CF37 5YR
(0443) 841888

Design Detectives

9 Greenside Road
London W12 9JQ
081-749 1489
Specializes in providing research relevant to design
projects.

Design Museum

Butlers Wharf
28 Shad Thames
London SE1 2YD
071-403 6933

Houses both permanent and changing collections of
contemporary and historic mass-produced designs.

Institute of Packaging (IOP)

Sysonby Lodge
Nottingham Road
Melton Mowbray
Leicestershire
LE13 0NU
(0664) 500055

London Chamber of Commerce

69 Cannon Street
London EC4N 5AB
071-248 4444

Market Research Society

15 Northburgh Street
London EC1V 0AH
071-490 4911

New Designers Exhibition

c/o The Business Design Centre
Upper Street
Islington Green
London N1 0QH
071-359 3535

Nigel French Enterprises

55 Colebrooke Row
London N1 8AF
071-354 8001
Offers forecasting services, eg on colour and styling trends.

Patent Office

Cardiff Road
Newport
Gwent NP9 1RH
(0633) 814000
Designs Enquiry Desk: Ext 5162
between 10.00 and 16.00 weekdays

Patent Office Search and Advisory Service

Hazlitt House
45 Southampton Buildings
Chancery Lane
London WC2A 1AR
071-438 4747

Peclers Agence de Style

23 Rue du Mail
75002 Paris
France
1 40 41 06 06
International forecasting agency, specializing in fashion,
textiles and household products.

Royal Society of Arts

8 John Adam Street
London WC2N 6EZ
071-930 5115

WHO Technical Report Series
875

TRAINING IN DIAGNOSTIC ULTRASOUND: ESSENTIALS, PRINCIPLES AND STANDARDS

2806174533

Report of a
WHO Study Group

World Health Organization
Geneva 1998

WHO Library Cataloguing in Publication Data

Training in diagnostic ultrasound : essentials, principles and standards : report of a
WHO study group

(WHO technical report series ; 875)

1.Ultrasonography 2.Diagnostic imaging — standards 3.Guidelines 4.Health per-
sonnel — education I.Title II.Series

ISBN 92 4 120875 9   (NLM Classification: WN 200)
ISSN 0512-3054

Typeset in Hong Kong
Printed in Spain
97/11650 — Best-set/Fotojae — 7500